Marriage and Family

Using MicroCase *ExplorIt*®

THIRD EDITION

Kevin Demmitt

CLAYTON COLLEGE & STATE UNIVERSITY

THOMSON

WADSWORTH

Australia • Canada • Mexico • Singapore • Spain • United Kingdom • United States

Publisher/Executive Editor: *Sabra Horne*
Acquisitions Editor: *Bob Jucha*
Technology Project Manager: *Julie Aguilar*
Assistant Editor: *Stephanie Monzon*
Editorial Assistant: *Matthew Goldsmith*
Marketing Manager: *Matthew Wright*

Marketing Assistant: *Michael Silverstein*
Printbuyer: *Karen Hunt*
Ancillary Coordinator: *Rita Jaramillo*
Printer: *Patterson Printing.*
Duplicator: *Micro Bytes, Inc.*

For more information about our products,
contact us at:
Thomson Learning Academic Resource Center
1-800-423-0563

For permission to use material from this text,
contact us by:
Phone: 1-800-730-2214
Fax: 1-800-731-2215
Web: www.thomsonrights.com

Asia
Thomson Learning
60 Albert Complex, #15-01
Albert Complex
Singapore 189969

Australia
Nelson Thomson Learning
102 Dodds Street
South Street
South Melbourne, Victoria 3205
Australia

Canada
Nelson Thomson Learning
1120 Birchmount Road
Toronto, Ontario M1K 5G4
Canada

Europe/Middle East/South Africa
Thomson Learning
Berkshire House
168-173 High Holborn
London WC1 V7AA
United Kingdom

Latin America
Thomson Learning
Seneca, 53
Colonia Polanco
11560 Mexico D.F.
Mexico

Spain
Paraninfo Thomson Learning
Calle/Magallanes, 25
28015 Madrid, Spain

CONTENTS

ABOUT THE AUTHOR

Kevin Demmitt is an Associate Professor of Sociology at Clayton College & State University near Atlanta, Georgia. He received his M.A. degree at Arizona State University and his Ph.D. from Purdue University. He has taught the Sociology of Marriage and Family course, and a wide range of other courses, at large state universities and smaller, liberal arts colleges throughout his career. His research and writing has focused on the relationship between family, work, and religion, but he has also researched topics related to politics, religion, and education. Professor Demmitt also conducts workshops on the effective use of computers in the classroom. He welcomes comments from students and faculty about this workbook. He can be reached by e-mail at KEVINDEMMITT@MAIL.CLAYTON.EDU or by regular mail at: Department of Social Sciences, Clayton College & State University, Morrow, GA 30260.

PREFACE

As you begin this marriage and family course, you bring to it a preconceived notion about family life. You have intuitive ideas about why some marriages seem to work while others fail. You have theories about the selection of mates and the raising of children. Whether you realize it or not, you have assumptions about the role of the family and its connection to society as a whole. But how do you know whether the ideas you hold are accurate, or even that your experiences and observations are a reliable reflection of society? That is where the sociology of marriage and the family comes in.

Through the use of empirical sociological research, it is possible to view the family from a broad, objective perspective. This workbook comes with five real data sets, each containing data used by sociologists in their own research. One of the data sets is based on surveys of the U.S. population, two data sets are drawn from cross-cultural sources, one data set is based on the fifty states, and a final data set allows the examination of historical trends related to family life. Quantitative research cannot tell us everything there is to know about the family. There are values and experiences related to marriage and family life that cannot be measured or quantified. But, by the time you finish this workbook you will discover that empirical research is one of the best tools for testing our ideas and assumptions about the social world.

WHAT'S NEW IN THE THIRD EDITION?

Similar to the previous editions, the data sets are extensive, current, and from the best sources available. This edition includes new data from the 2000 General Social Survey and the 2000 U.S. Census, as well as updated variables in the contemporary cross-cultural data set. One of the themes that is highlighted throughout this third edition is the relationship between personal choice and social forces. Personal choice helps account for the diversity of family structures and lifestyles in the United States and cross-culturally, but the choices people make are influenced by the larger social environment. Finally, this third edition introduces a new family life stage variable to illustrate how people may behave or think differently depending upon their current family situation.

ACKNOWLEDGMENTS

The writing and development of this book was in many ways a team project. I would especially like to thank those at Wadsworth who provided direction, encouragement, and innovative ideas for this third edition: Eve Howard, Editor in Chief of Behavioral and Social Sciences, Analie Barnett, Assistant Editor, and Bob Jucha, Acquisitions Editor for Sociology. Their personal touches and expertise are evident throughout the book. And, Executive Sales Representative Davene Staples has kept me informed on current trends in marriage and family textbooks. MicroCase Technology Program Manager Julie Aguilar's insight into the pedagogical and technical features of this book was invaluable. Jodi Gleason once again guided this project through the production and editorial stages with professionalism and great efficiency. And it was a pleasure to work with the MicroCase data archivist, Matt Bahr. I also continue to be appreciative to David Smetters, who was instrumental in developing the first two editions of this book.

I would like to thank the original sources of the data files that accompany this workbook. Special thanks goes to Tom W. Smith at the National Opinion Research Center for his continued direction and administration of the General Social Survey. Many variables in the global data set are based on the World Values Survey, for which we must thank Ronald Inglehart at the Institute for Social Research, University of Michigan. Thanks also goes to Child Trends, Inc., which maintains and distributes the National Survey of Children for the American Family Data Archive at Sociometrics Corporation.

I would also like to thank the reviewers of the third edition, Mary Kirby Diaz at SUNY-Farmingdale, Norval D. Glenn at the University of Texas at Austin, Nita Jackson at Butler County Community College, Christine Monnier at the College of DuPage, and Gilbert Zicklin at Montclair State University. I continue to be appreciative of the reviewers of the first two editions: Stephen Bahr at Brigham Young University, Elaine Wethington at Cornell University, and Ronaele Whittington at The University of Hawaii-Manoa, Ray Darville at Stephen F. Austin University, Jerry Michel at the University of Memphis, Richard Miller at Missouri Southern State College, Charles O'Connor at Bemidji State University, Leland Robinson at the University of Tennessee-Chattanooga, Ed Sabin at Towson State University, and Roberta Satow at Brooklyn College.

I also wish to acknowledge my friends and colleagues at Clayton College & State University. Their commitment to innovative and effective teaching is a source of inspiration for trying new ideas in the classroom.

Lastly, I wish to express my gratitude to my wife, Audrey; my three children, Andrew, Hannah, and Jacob; my parents, and my four brothers. They are the ones who have taught me the value and joy of family life.

GETTING STARTED

INTRODUCTION

Welcome to ExplorIt! With the easy-to-use software accompanying this workbook, you will have the opportunity to learn about marriage and family by using some of the same data that researchers use. The goal of this workbook is to help you learn how to use data to explore marriage and the family from a sociological perspective and how to investigate new ideas and conduct research to test these ideas.

Each chapter in this workbook has two sections. The first section introduces a particular area of marriage and family research and demonstrates how data are used to support, augment, and test the ideas proposed. Following the introduction is the worksheet section. The worksheets begin with a few review questions on the material covered in the preliminary section. Then, additional data analysis exercises are provided that further explore the topics under investigation in the chapter. You will use the student version of ExplorIt to complete these worksheets. You can easily create all the graphics by following the ExplorIt Guides you'll be seeing. Doing so will take just a few clicks of your computer mouse and will help you become familiar with ExplorIt. The ExplorIt Guides are described in more detail below.

SYSTEM REQUIREMENTS

- Windows 95 (or higher)
- 8 MB RAM
- CD-ROM drive
- 15 MB of hard drive space (if you want to install it)

To run the software on a Macintosh, you will need emulation software or hardware installed. For more information about emulation software or hardware, check with your local Macintosh retailer or try the Web-site http://machardware.about.com/cs/pcemulation/.

NETWORK VERSIONS OF STUDENT EXPLORIT

A network version of Student ExplorIt is available at no charge to instructors who adopt this book for their course. It's worth noting that Student ExplorIt can be run directly from the CD-ROM on virtually any computer network—regardless of whether a network version of Student ExplorIt has been installed.

INSTALLING STUDENT EXPLORIT

If you will be running Student ExplorIt directly from the CD-ROM—or if you will be using a version of Student ExplorIt that is installed on a network—skip to the section "Starting Student ExplorIt."

To install Student ExplorIt to a hard drive, you will need the CD-ROM that is packaged inside the back cover of this book. Then follow these steps in order:

1. Start your computer and wait until the Windows desktop is showing on your screen.

2. Insert the CD-ROM disc into the CD-ROM drive of your computer.

3. On most computers the CD-ROM will automatically start and a welcome menu will appear. If the CD-ROM doesn't automatically start, do the following:

   Click [Start] from the Windows desktop, click [Run], type **D:\SETUP**, and click [OK]. (If your CD-ROM drive is not the D drive, replace the letter D with the proper drive letter.) To install Student ExplorIt to your hard drive, select the second option on the list: "Install Student ExplorIt to your hard drive."

4. During the installation, you will be presented with several screens, as described below. In most cases you will be required to make a selection or entry and then click [Next] to continue.

   The first screen that appears is the **License Name** screen. Here you are asked to type your name. It is important to type your name correctly, since it cannot be changed after this point. Your name will appear on all printouts, so make sure you spell it completely and correctly! Then click [Next] to continue.

   A **Welcome** screen now appears. This provides some introductory information and suggests that you shut down any other programs that may be running. Click [Next] to continue.

   You are next presented with a **Software License Agreement**. Read this screen and click [Yes] if you accept the terms of the software license.

   The next screen has you **Choose the Destination** for the program files. You are strongly advised to use the destination directory that is shown on the screen. Click [Next] to continue.

5. The Student ExplorIt program will now be installed. At the end of the installation, you will be asked if you would like a shortcut icon placed on the Windows desktop. We recommend that you select [Yes]. You are now informed that the installation of Student ExplorIt is finished. Click the [Finish] button and you will be returned to the opening Welcome screen. To exit completely, click the option "Exit Welcome Screen."

STARTING STUDENT EXPLORIT

There are three ways to run Student ExplorIt: (1) directly from the CD-ROM, (2) from a hard drive installation, or (3) from a network installation. Each method is described below.

Starting Student ExplorIt from the CD-ROM

Unlike most Windows programs, it is possible to run Student ExplorIt directly from the CD-ROM. To do so, follow these steps:

1. Insert the CD-ROM disc into the CD-ROM drive.

2. On most computers the CD-ROM will automatically start and a Welcome menu will appear. (Note: If the CD-ROM does **not** automatically start after it is inserted, click [Start] from the Windows desktop, click [Run], type **D:\SETUP**, and click [OK]. If your CD-ROM drive is not the D drive, replace the letter D with the proper drive letter.)

3. Select the first option from the Welcome menu: **Run Student ExplorIt from the CD-ROM**. Within a few seconds, Student ExplorIt will appear on your screen. Type in your name where indicated to enter the program.

Starting Student ExplorIt from a Hard Drive Installation

If Student ExplorIt is installed to the hard drive of your computer (see earlier section "Installing Student ExplorIt"), it is **not** necessary to insert the CD-ROM. Instead, locate the Student ExplorIt "shortcut" icon on the Windows desktop, which looks something like this:

To start Student ExplorIt, position your mouse pointer over the shortcut icon and double-click (that is, click it twice in rapid succession). If you did not permit the shortcut icon to be placed on the desktop during the install process (or if the icon was accidentally deleted), you can alternatively follow these directions to start the software:

Click [Start] from the Windows desktop.

Click [Programs].

Click [MicroCase].

Click [Student ExplorIt - MA].

After a few seconds, Student ExplorIt will appear on your screen.

Starting Student ExplorIt from a Network

If the network version of Student ExplorIt has been installed to a computer network, you need to double-click the Student ExplorIt icon that appears on the Windows desktop to start the program. Type in your name where indicated to enter the program. (Note: Your instructor may provide additional information that is unique to your computer network.)

MAIN MENU OF STUDENT EXPLORIT

Student ExplorIt is extremely easy to use. All you do is point and click your way through the program. That is, use your mouse arrow to point at the selection you want, then click the left button on the mouse.

The main menu is the starting point for everything you will do in Student ExplorIt. Look at how it works. Notice that not all options on the menu are always available. You will know which options are available at any given time by looking at the colors of the options. For example, when you first start the software, only the OPEN FILE option is immediately available. As you can see, the colors for this option are brighter than those for the other tasks shown on the screen. Also, when you move your mouse pointer over this option, it is highlighted.

EXPLORIT GUIDES

Throughout this workbook, "ExplorIt Guides" provide the basic information needed to carry out each task. Here is an example:

> ➤ *Data File:* **STATES**
> ➤ *Task:* **Mapping**
> ➤ *Variable 1:* **3) %URBAN**
> ➤ *View:* **Map**

Each line of the ExplorIt Guide is actually an instruction. Let's follow the simple steps to carry out this task.

Step 1: Select a Data File

Before you can do anything in Student ExplorIt, you need to open a data file. To open a data file, click the OPEN FILE task. A list of data files will appear in a window (e.g., GSS, STATES). If you click on a file name *once*, a description of the highlighted file is shown in the window next to this list. In the ExplorIt Guide shown above, the ➤ symbol to the left of the Data File step indicates that you should open the STATES data file. To do so, click STATES and then click the [Open] button (or just double-click STATES). The next window that appears (labeled File Settings) provides additional information about the data file, including a file description, the number of cases in the file, and the number of variables, among other things. To continue, click the [OK] button. You are now returned to the main menu of Student ExplorIt. (You won't need to repeat this step until you want to open a different data file.) Notice that you can always see which data file is currently open by looking at the file name shown on the top line of the screen.

Step 2: Select a Task

Once you open a data file, the next step is to select a program task. Eight analysis tasks are offered in this version of Student ExplorIt. Not all tasks are available for each data file, because some tasks are appropriate only for certain kinds of data. Mapping, for example, is a task that applies only to ecological data, and thus cannot be used with survey data files.

In the ExplorIt Guide we are following, the ➤ symbol on the second line indicates that the MAPPING task should be selected, so click the MAPPING option with your left mouse button.

Step 3: Select a Variable

After a task is selected, you will be shown a list of the variables in the open data file. Notice that the first variable is highlighted and a description of that variable is shown in the Variable Description window at the lower right. You can move this highlight through the list of variables by using the up and down cursor keys (as well as the <Page Up> and <Page Down> keys). You can also click once on a variable name to move the highlight and update the variable description. Go ahead—move the highlight to a few other variables and read their descriptions.

If the variable you want to select is not showing in the variable window, click on the scroll bars located on the right side of the variable list window to move through the list. See the following figure.

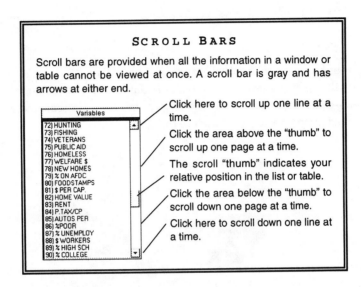

By the way, you will find an appendix at the back of this workbook that contains a list of the variable names for key data files provided in this package.

Each task requires the selection of one or more variables, and the ExplorIt Guides indicate which variables you should select. The ExplorIt Guide example here indicates that you should select 3) %URBAN as Variable 1. On the screen, there is a box labeled Variable 1. Inside this box, there is a vertical cursor that indicates that this box is currently an active option. When you select a variable, it will be placed in this box. Before selecting a variable, be sure that the cursor is in the appropriate box. If it is not, place the cursor inside the appropriate box by clicking the box with your mouse. This is important because in some tasks the ExplorIt Guide will require more than one variable to be selected, and you want to be sure that you put each selected variable in the right place.

To select a variable, use any one of the methods shown below. (Note: If the name of a previously selected variable is in the box, use the <Delete> or <Backspace> key to remove it—or click the [Clear All] button.)

- Type the **number** of the variable and press <Enter>.

- Type the **name** of the variable and press <Enter>. Or you can type just enough of the name to distinguish it from other variables in the data—URB would be sufficient for this example.

- Double-click the desired variable in the variable list window. This selection will then appear in the variable selection box. (If the name of a previously selected variable is in the box, the newly selected variable will replace it.)

- Highlight the desired variable in the variable list, then click the arrow that appears to the left of the variable selection box. The variable you selected will now appear in the box. (If the name of a previously selected variable is in the box, the newly selected variable will replace it.)

Once you have selected your variable (or variables), click the [OK] button to continue to the final results screen.

Step 4: Select a View

The next screen that appears shows the final results of your analysis. In most cases, the screen that first appears matches the "view" indicated in the ExplorIt Guide. In this example, you are instructed to look at the Map view—that's what is currently showing on the screen. In some instances, however, you may need to make an additional selection to produce the desired screen.

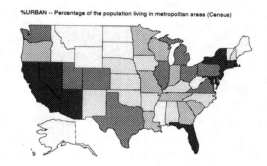
%URBAN -- Percentage of the population living in metropolitan areas (Census)

(OPTIONAL) Step 5: Select an Additional Display

Some ExplorIt Guides will indicate that an additional "Display" should be selected. In that case, simply click on the option indicated for that additional display. For example, this ExplorIt Guide may have included an additional line that required you to select the Legend display.

Step 6: Continuing to the Next ExplorIt Guide

Some instructions in the ExplorIt Guide may be the same for at least two examples in a row. For instance, after you display the map for population in the example above, the following ExplorIt Guide may be given:

> Data File: **STATES**
> Task: **Mapping**
> ➤ Variable 1: **6) POP.GROW**
> ➤ View: **Map**

Notice that the first two lines in the ExplorIt Guide do not have the ➤ symbol in front of the items. That's because you already have the data file STATES open and you have already selected the MAPPING task. With the results of your first analysis showing on the screen, there is no need to return to the main menu to complete this next analysis. Instead, all you need to do is select POP.GROW as your new variable. Click the [[⟲]] button located in the top left corner of your screen and the variable selection screen for the MAPPING task appears again. Replace the variable with 6) POP.GROW and click [OK].

To repeat: You need to do only those items in the ExplorIt Guide that have the ➤ symbol in front of them. If you start from the top of the ExplorIt Guide, you're simply wasting your time.

If the ExplorIt Guide instructs you to select an entirely new task or data file, you will need to return to the main menu. To return to the main menu, simply click the [Menu] button at the top left corner of the screen. At this point, select the new data file and/or task that is indicated in the ExplorIt Guide.

That's all there is to the basic operation of Student ExplorIt. Just follow the instructions given in the ExplorIt Guide and point and click your way through the program.

ONLINE HELP

Student ExplorIt offers extensive online help. You can obtain task-specific help by pressing <F1> at any point in the program. For example, if you are performing a scatterplot analysis, you can press <F1> to see the help for the SCATTERPLOT task.

If you prefer to browse through a list of the available help topics, select **Help** from the pull-down menu at the top of the screen and select the **Help Topics** option. At this point, you will be provided a list of topic areas. Each topic is represented by a closed-book icon. To see what information is available in a given topic area, double-click on a book to "open" it. (For this version of the software, use only the "Student ExplorIt" section of help; do not use the "Student MicroCase" section.) When you double-click on a book graphic, a list of help topics is shown. A help topic is represented by a graphic with a piece of paper with a question mark on it. Double-click on a help topic to view it.

If you have questions about Student ExplorIt, try the online help described above. If you are not very familiar with software or computers, you may want to ask a classmate or your instructor for assistance.

EXITING FROM STUDENT EXPLORIT

If you are continuing to the next section of this workbook, it is *not* necessary to exit from Student ExplorIt quite yet. But when you are finished using the program, it is very important that you properly exit the software—do not just walk away from the computer or remove your CD-ROM. To exit Student ExplorIt, return to the main menu and select the [Exit Program] button that appears on the screen.

Important: If you inserted your CD-ROM before starting Student ExplorIt, remember to remove it before leaving the computer.

PERSONAL CHOICES AND SOCIAL FORCES

Tasks: Mapping, Scatterplot, Correlation, Historical Trends
Data Files: STATES, TRENDS

Marriage, as a social institution, is characterized by the seemingly contradictory features of both stability and change. The impetus for change comes from the relative freedom of choice Americans enjoy when it comes to personal relationships. Unlike in other times or places where marriages might be arranged by parents or others in the community, we have the freedom to decide not only who and when we will marry, but whether we will marry at all. With so much freedom of choice, you might think that the structure of marriage and family life in our society would be completely unpredictable. It would be impossible to guess how many people will get married in a given year or in which part of the country the greatest number of people will get divorced. And, in fact, trying to predict familial outcomes for any individual is extremely difficult. But, if we step back a bit and look at the bigger picture, some trends and patterns do begin to emerge. As we will see throughout this book, there is a connection between those larger trends and patterns and the decisions made by individuals.

Contrary to what one might think, freedom of choice is not incompatible with stability. Take the game of baseball, for instance. In every game, individual decisions are constantly being made that could determine the final outcome of the game. With each of the hundreds of pitches thrown in a game, the pitcher decides where and how to throw the ball and the batter decides whether or not to swing. With so many variables in play, one might think there would be huge variations as to how well one team hits the ball when compared to another. Yet, the actual difference in hitting between major league teams over the course of a season is incredibly small. For example, the 2000 World Series champion New York Yankees team batting average for the season was only 2 percentage points higher than the worst hitting team in baseball that year. The Yankees hit safely 27.7 percent of the time compared to 25.7 percent of the time for the Devil Rays. When examined over the length of an entire season, there is incredible stability when it comes to the likelihood of players hitting safely. Despite this stability, each at bat can be exciting for a baseball fan because the outcome of any given pitch is unknown. The same can be said for the study of marriage and the family. We never know what each day might bring for an individual, but we can study trends related to the larger institutions of marriage and the family. We can also try to understand why people make different choices and what the outcomes of those choices are likely to be.

The family, like all other social institutions, can be viewed through a microscope or a telescope. For example, if I asked you to describe *your* family, you would give me the microscopic view. You would probably describe your parents, siblings, and other relatives, and the experiences you had growing up. However, if I asked you to write an essay on *the* family, you would undoubtedly give me a very different answer. Rather than describing unique individuals, you would focus on social positions, or **statuses**, such as those of mothers, fathers, children, grandparents, husbands, wives, and so on, and the social expectations, or **roles**, attached to each status. The combination of social statuses that make up a family unit is referred to as the **family structure**. In looking at the family from this perspective, we see an institution that is characterized by diversity. In addition to the **nuclear family**, consisting of father, mother, and children, there are single-parent families, families with stepchildren, couples without children, multigenerational families, and a host of other family structures. In this chapter, we will examine the diversity of family structures across the United States and explore the relationship between personal choices and the larger social environment. Along the way, you will learn how to operate the accompanying software and master some basic research techniques.

Does the place where you live have any bearing on whether or not you will get married? Are you more likely to get divorced if you happen to live in the West than if you live in the South or the Midwest? Many people would answer *no* to both of these questions. After all, getting married or divorced is an individual decision—how could the region of the country you live in possibly make a difference? On the other hand, if regions of the country vary with regard to their cultures or economic characteristics, then where you live *could* make a difference. Perhaps people who live in one place share family-related values that differ from those of people who live elsewhere. Maybe the availability of good jobs or schools, or a number of other environmental factors that vary geographically, have an impact on the familial decisions people make. If this is the case, we would in fact expect to find regional differences in family structures.

Let's use the data we have available to see whether family structures are the same or different from one region of the United States to another. In a standard textbook, such a comparison would be dependent on printed maps showing the regional distribution of various family structures. In fact, such maps appear on the following pages in this book. But what makes this workbook different from an ordinary textbook is that *you* are able to generate a map of *any* item included in the STATES data file—whether it is a map of the divorce rates or the average SAT test scores in the 50 states. And, once a map is on your computer screen, you can do a lot of other things to it, as you will discover as you proceed through this chapter.

To begin, let's take a look at the percentage of households occupied by married couples to see if there are regional variations.

➤ *Data File:* **STATES**
➤ *Task:* **Mapping**
➤ *Variable 1:* **22) %MARRIED**
➤ *View:* **Map**

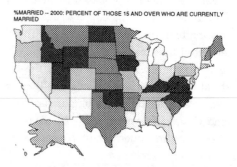

%MARRIED -- 2000: PERCENT OF THOSE 15 AND OVER WHO ARE CURRENTLY MARRIED

To reproduce this graphic on the computer screen using ExplorIt, review the instructions in the "Getting Started" section. For this example, you would open the STATES data file, select the MAPPING task, and select 22) %MARRIED for Variable 1. The first view shown is the Map view. (Remember, the ➤ symbol indicates which steps you need to perform if you

are doing all the examples as you follow along in the text.) So in the next example below, you only need to select a new view—that is, you don't need to repeat the first three steps because they were already done in this example.

In this map of the United States, the states appear in several colors from very dark to very light based on the percentage of households occupied by married couples according to the U.S. Census. The darker a state looks, the greater its percentage of married couples. The data for married couples, as with most of the variables in the STATES data set, is presented in the form of a percentage. Thus, the relative size of a state's population does not distort the results.

Indeed, there are some distinct regional variations evident in this map. The distribution of the married population in the United States is not random. Most of the states with the highest couple rates are clustered in the central portion of the country. The states with the lowest couple rates tend to be found along the coasts. Now let's look at the actual percentages.

%MARRIED: Percent of those 15 and over who are currently married

Data File: **STATES**
Task: **Mapping**
Variable 1: **22) %MARRIED**
➤ View: **List: Rank**

RANK	CASE NAME	VALUE
1	Utah	60.70
2	North Dakota	60.16
3	North Carolina	60.06
4	Kentucky	59.94
5	Idaho	59.75
6	Iowa	59.73
7	Oklahoma	59.51
8	West Virginia	59.30
9	Wyoming	59.04
10	Virginia	59.03

As indicated by the ➤ symbol, if you are continuing from the previous example, select the [List: Rank] option. The number of rows shown on your screen may be different from that shown here. Use the cursor keys and scroll bar to move through this list if necessary.

Utah has the largest percentage of married couples, with 60.70 percent of its households being occupied by married couples in 2000. North Dakota is number two (60.16 percent). Moving down the list we see that five of the ten states with the most married couples are located in the north central United States. New York is in last place, with 53.14 percent. Rhode Island ranks second from the bottom, with 53.17 percent.

The percentage of households occupied by a married couple is an example of what social researchers call a *variable*. A **variable** is anything that varies among the objects being examined. For example, all states have rules as to what constitutes a marriage. Therefore, states that have legally defined marriages *is not* a variable because all states have legally defined marriages. However, the number of legally recognized marriages in each state does vary. Thus, the *percentage* of the population in each state that is married is a variable. States also differ in the proportions of their populations who remain single, have children, go camping, go to church, are wealthy, or drop out of school—all of these could be variables. Or, if we are examining individuals rather than states, all traits and characteristics in which people differ—height, weight, political opinions, hobbies, or education, for example—are variables.

The basic task of social science is to *explain variation*. We do this by trying to *discover connections among variables*. Now that we have seen that the proportion of married couples varies by region in this country, the next step is to try to explain *why* there are fewer married couples outside of America's heartland. Is it because people in other parts of the country are less likely to marry? Or could it be that the likelihood of getting married is the same everywhere, but that couples in some parts of the country are more likely to be separated or divorced? Either of these factors, singleness or divorce, could affect the percentage of married couples. To see whether one or both of these factors influence the married-couples rate, we need to see where the greatest percentage of divorced and single persons is located.

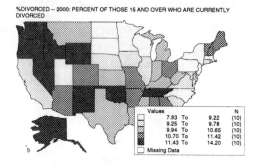

Data File: **STATES**
Task: **Mapping**
➤ Variable 1: **14) %DIVORCED**
➤ View: **Map**
➤ Display: **Legend**

When the map appears, click the [Legend] option to display the legend.

This map presents a snapshot of persons who are currently divorced (those who have divorced and remarried were represented in the map of married couples). Whereas singleness is a bicoastal phenomenon, divorce is skewed more to the west. Most of the states with percentages of divorced residents in the 9.9 to 14 percent range lie west of the Mississippi River.

%DIVORCED: Percent of those 15 and over who are currently divorced

Data File: **STATES**
Task: **Mapping**
Variable 1: **14) %DIVORCED**
➤ View: **List: Rank**

RANK	CASE NAME	VALUE
1	Nevada	14.20
2	Wyoming	12.58
3	Oregon	12.31
4	Florida	11.93
5	New Mexico	11.91
6	Oklahoma	11.84
7	Idaho	11.76
8	Washington	11.68
9	Tennessee	11.67
10	Alaska	11.43

Nevada has the highest rate of residents who are currently divorced, with 14.2 percent, followed by Wyoming (12.58 percent) and Oregon (12.31 percent). New Jersey (7.83 percent) has the lowest rate of divorced residents. The north central states are among those with the fewest divorced residents. So one of the factors that explains the high couples rate in the central states is that there are fewer divorced residents living there. Now let's look at the distribution of single adults to see where they are concentrated geographically.

Data File: **STATES**
Task: **Mapping**
➤ Variable 1: **23) %SINGLE**
➤ View: **Map**

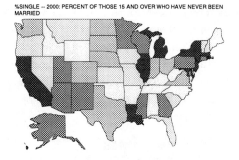

%SINGLE -- 2000: PERCENT OF THOSE 15 AND OVER WHO HAVE NEVER BEEN MARRIED

If you are continuing from the previous example, return to the variable selection screen for the MAPPING task, then select 23) %SINGLE as Variable 1. (It is not necessary to reselect the STATES data file or the MAPPING task.)

Here we see the percentage of the population in each state who are 15 and older and have never been married. This map is nearly a reverse image of the map for married couples. The coastal states are generally darker than the central states, which means they have a greater percentage of single adults. Let's take a look at the exact percentages.

Data File: **STATES**
Task: **Mapping**
Variable 1: **23) %SINGLE**
➤ View: **List: Rank**

%SINGLE: Percent of those 15 and over who have never been married

RANK	CASE NAME	VALUE
1	New York	31.52
2	California	30.69
3	Massachusetts	30.07
4	Rhode Island	30.01
5	Maryland	29.40
6	Illinois	29.28
7	Hawaii	28.63
8	Louisiana	28.63
9	New Jersey	28.49
10	Michigan	28.34

New York (31.52 percent) and California (30.69 percent) have the greatest percentage of adults who have never been married. Oklahoma has the lowest percentage of adults who have never married (21.47 percent), followed by West Virginia, Arkansas, and Kentucky. Many of the states that had a high percentage of married couples are clustered among the states that have a low percentage of unmarried adults. So not only are there fewer divorced individuals living in America's heartland, there are fewer single adults living there as well. Lower rates of both singleness and divorce help to explain the high percentage of married couples in the central states. People who live there are more likely to marry and also are more likely to stay married.

Later in this chapter we will look at some more variations in family structures, but let's explore this distribution of the singles population a little further. In the United States, deciding if and when to get married is an individual choice. Unlike some other cultures, we don't have arranged marriages whereby parents choose a spouse for their child based upon social or economic considerations. So, *why* is there a lower percentage of people who have made the choice to get married living along the coasts than in the central states? In other words, how might the larger social environment influence such a personal decision?

Chapter 1: Personal Choices and Social Forces

One way the social structure affects individual decisions is by controlling the options from which an individual can choose. For example, one possible explanation for the larger number of singles in some states than in others would be an imbalance in the gender ratio of available partners. In other words, there may be an overabundance of single men or single women along the coasts so that there simply aren't enough potential marriage partners to go around. This could be caused by migration or mortality patterns that increase or decrease the population of one sex more than the other. To see whether that is the case, we need to compare a map of single males with a map of single females.

Data File: **STATES**
Task: **Mapping**
➤ Variable 1: **15) %SINGLE M**
➤ Variable 2: **18) %SINGLE F**
➤ Views: **Map**

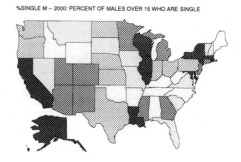

%SINGLE M -- 2000: PERCENT OF MALES OVER 15 WHO ARE SINGLE

r = 0.896**

%SINGLE F -- 2000: PERCENT OF FEMALES OVER 15 WHO ARE SINGLE

If you are continuing from the previous example, return to the variable selection screen for the MAPPING task. Select 15) %SINGLE M for Variable 1 and 18) %SINGLE F for Variable 2.

The map representing the distribution of single males across the country appears on the top half of the screen. The map showing the distribution of the single female population appears on the bottom half of the screen.

The two maps look surprisingly similar. The highest concentrations of single males are located in the same regions as the highest concentrations of single females. Both single men and single women are most likely to live in the Northeast or along the west coast. The lower married-couples rates in these areas cannot be attributed to a lack of available marriage partners.

If there are plenty of potential marriage partners along the coast, then why is it that fewer people who live there are choosing to marry? Perhaps it is that the cultures in these areas offer single individuals more options. For numerous reasons, including greater racial and ethnic diversity of cities, people who live in urban areas also tend to be more tolerant of diverse lifestyles, so there may be less pressure on individuals to marry. Cities also have more career and entertainment opportunities that compete with marriage. Could it be that the coastal states have more singles because they have larger urban populations? Let's compare maps of the percentage of singles with the percentage of the population living in urban areas.

Data File: **STATES**
Task: **Mapping**
➤ Variable 1: **23) %SINGLE**
➤ Variable 2: **3) %URBAN**
➤ Views: **Map**

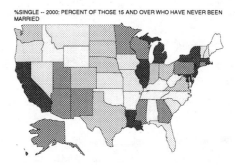

%SINGLE -- 2000: PERCENT OF THOSE 15 AND OVER WHO HAVE NEVER BEEN MARRIED

r = 0.613**

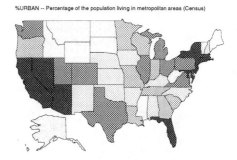

%URBAN -- Percentage of the population living in metropolitan areas (Census)

These maps look very similar to each other. It appears that states where most of the population live in urban areas tend to offer more alternatives to marriage. Although individuals across all 50 states have the freedom to choose if and when to marry, the social structure of urban areas offers options that make marriage less likely. Individuals have free choice, but the choices people make are somewhat predictable depending on the characteristics of the surrounding social environment.

What we have seen thus far demonstrates that family structures are not randomly distributed throughout the country. Although nearly half of the adults living in every state are married, there are some distinct trends. The states along the coasts have more singles. There are more divorced persons in the West. The north central states have the highest percentage of married couples. These are all trends that you would miss if you looked only at isolated individuals.

Although most people consider getting married or divorced (or a similar familial decision) to be an individual choice, the trends we have observed tell us that these decisions are influenced by factors beyond the individual. The larger sociocultural environment does play a role in the decisions people make.

Comparing maps is one way to look for relationships between variables, but it is somewhat subjective. For example, the last two maps you compared were similar but not identical. Some states that had more urban residents did not have more singles. It also becomes more difficult to say how similar any two maps are when the maps are very complex. For example, it would be much harder to compare two maps based on the 3,142 counties of the United States than two maps based on the 50 states. The same is true of attempts to compare lists. It is not hard to compare lists of the 50 states, but it would be much more difficult to compare longer lists. Thus, it was a considerable achievement when, in the 1890s, an Englishman named Karl Pearson discovered an extremely simple method for comparing maps or lists.

To see Pearson's method, we can draw a horizontal line across the bottom of a sheet of paper. We will let this line represent the size of the singles population. At the left end of this line, we will write 21.47, which indicates the state with the smallest proportion of single adults: Oklahoma. At the right end of the line, we will place the number 31.52 to represent New York, the state with the largest proportion of singles.

21.47	31.52

Now we can draw a vertical line up the left side of the paper. This line will represent the proportion of urban residents in a given state. At the bottom of this line, we will write 23.8 to represent Montana, the state with the smallest urban population. At the top, we will write 100 to represent New Jersey, the most urban state.

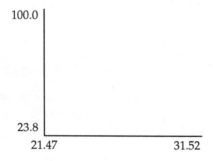

Now that we have a line with an appropriate scale to represent each map, the next thing we need to do is look at the distribution for each map to obtain the value for each state and then locate it on each side according to its score. Let's start with New Jersey. Since it has the largest urban population, we can easily find its place on the horizontal axis above. Make a small mark at 100 to locate New Jersey. Next, New Jersey has the ninth highest singles rate, so estimate where a rate of 28.49 is located on the vertical line. Knowing where New Jersey is on each line representing each map, we draw a vertical line up from its position on the line for urban residents and we draw a horizontal line out from its position on the line for single adults. Where these two lines meet (or cross), we make a dot. This dot represents the combined map location of New Jersey.

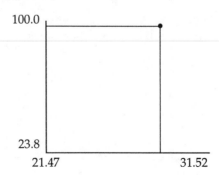

Now let's locate California. Its proportion of urban residents is 96.7, so we make a mark on the horizontal line at that spot. Its singles rate is 30.69, so we mark that point on the vertical line. The point where these two lines meet is the combined map location for California. When we have followed this procedure for each state, we will have 50 dots located within the space defined by the vertical and horizontal lines representing the two maps. What we have done is create a **scatterplot**. Fortunately, you don't have to go to all this trouble. ExplorIt will do it for you.

Data File: **STATES**

➤ Task: **Scatterplot**

➤ Dependent Variable: **23) %SINGLE**

➤ Independent Variable: **3) %URBAN**

➤ View: **Reg. Line**

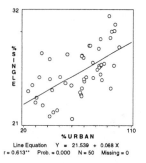

Notice that the scatterplot requires two variables.

Special feature: When the scatterplot is showing, you may obtain the information on any dot by clicking on it. A little box will appear around the dot, and the values of 3) URBAN (or the X-axis variable) and of 23) %SINGLE (or the Y-axis variable) will be shown.

Each of these dots represents a state. Which state is represented by the dot at the top of the screen? That's right—New York. It has the highest singles rate and one of the largest urban populations. When you click on the dot for New York, additional information about that state will be provided to the left of the scatterplot. The "X" value of 91.7 indicates New York's value on the X-axis (the horizontal axis representing the proportion of urban residents). The "Y" value of 31.52 indicates New York's value on the "Y" axis (the vertical axis representing the singles rate).

Once Pearson had created a scatterplot, his next step was to calculate what he called the *regression line*—which you can see running diagonally across the scatterplot. This line represents the best effort to draw a straight line that connects all the dots. It is unnecessary for you to know how to calculate the location of the regression line because the program does it for you. But if you would like to see how the regression line would look if all the dots were located along a straight line, all you need to do is examine the scatterplot for identical maps. So, if you create a scatterplot using %SINGLE for both the dependent and independent variables, you will be comparing identical maps and the dots representing states will all be on the regression line like a straight string of beads.

However, since the maps for urban population and singles are only very similar, but not identical, most of the dots are scattered near, but not on, the regression line. Pearson's method for calculating how similar any two maps or lists are becomes very easy once the regression line has been drawn. What it amounts to is measuring the distance out from the regression line to every dot.

Data File: **STATES**
Task: **Scatterplot**
Dependent Variable: **23) % SINGLE**
Independent Variable: **3) %URBAN**
➤ *View:* **Reg. Line/Residuals**

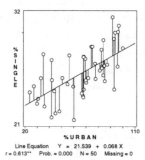

Line Equation   Y = 21.539 + 0.068 X
r = 0.613**   Prob. = 0.000   N = 50   Missing = 0

> **To show the residuals, select the [Residuals] option. Keep the [Reg. Line] option selected as well.**

See all the little lines. If you added them all together, you would have the sum of the deviations of the dots from the regression line. The smaller this sum, the more similar are the two maps. For example, when the maps are identical and all the dots are on the regression line, the sum of the deviations is zero.

To make it simple to interpret results, Pearson invented a procedure to convert the sums into a number he called the **correlation coefficient**. The correlation coefficient varies from 0.0 to 1.0. When the maps are identical, the correlation coefficient will be 1.0. When the maps are completely different, the correlation coefficient will be 0.0. Thus, the closer the correlation coefficient is to 1.0, the more alike the two maps or lists are. Pearson used the letter r as the symbol for his correlation coefficient.

Look at the lower left of the screen above and you will see r = 0.613. (The meaning of the asterisks will be explained a bit later.) This indicates that the maps are similar. There are no steadfast rules for interpreting the strength of a correlation coefficient (r), but as a general guideline, the following rule of thumb can be used with the STATES data file:

.70 or higher	Very strong relationship
.40 to .69	Strong relationship
.30 to .39	Moderate relationship
.20 to .29	Weak relationship
below .20	No or negligible relationship

Keep in mind that correlation and causation are *not* the same thing. It is true that without correlation there can be no causation. Thus, the urban population rate cannot be considered a contributing factor to the singles rate *if* there is no correlation between the two variables. But correlations often occur between two variables without one causing the other. For example, in any grade school you would find a very high correlation between children's height and their reading ability. However, as we know, being taller does not really affect a child's ability to read. This correlation occurs because both height and reading ability reflect age—the taller kids are older and the older kids read better. The positive correlation between the singles rate and the proportion of the population living in cities may signify a cause-and-effect relationship. However, we cannot rule out the possibility that this positive correlation may actually reflect some other characteristic of the coastal urban areas. For example, urban areas experience more population growth than rural areas. Perhaps the high singles rate in urban areas has nothing to do with the culture of city life but it is just the result of cities attracting single adults from other regions of the country who simply have not yet had the opportunity to meet someone to marry.

Data File: **STATES**
Task: **Scatterplot**
Dependent Variable: **23) %SINGLE**
➤ Independent Variable: **6) POP.GROW**
➤ View: **Reg. Line**

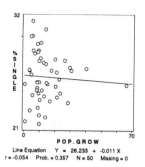

Line Equation   Y = 26.233 + -0.011 X
r = -0.054   Prob. = 0.357   N = 50   Missing = 0

This scatterplot illustrates the relationship between the percentage of singles and the population growth rate. Do areas with more rapid population growth have more single adults? No. This is what a scatterplot looks like when two variables are not correlated. The dots are scattered all over the screen. The regression line has only a slight slope as it crosses the screen from left to right. However, the value of r is not zero (r = –0.054). So, how can we say these two variables are not correlated? We can say it because the odds are very high that this correlation is nothing but a random accident.

In Chapter 2 you will learn how social scientists calculate the odds as to whether or not a correlation is random. Here it is sufficient to know that many correlations are so small, we treat them *as if they were zero*. And the software automatically does that calculation for you and gives you the results. If you look back at the correlation between singles and urbanness, you will see that there are two asterisks following the value of r (r = .613**). Two asterisks means that there is less than 1 chance in 100 that this correlation is a random accident. One asterisk means that the odds against a correlation being random are 20 to 1. Whenever there are no asterisks following a correlation, the odds are too high that it could be random.[1] That's how we know that the correlation between population growth and the singles rate is too small too matter—there are no asterisks. *Treat all correlations without asterisks as zero correlations.*

Remember that correlation does not necessarily demonstrate causation; variables may be highly correlated without one having caused the other. Nevertheless, social scientists examine correlations primarily to test hypotheses that propose cause-and-effect relationships. To more fully capture this aspect of research, it is helpful to distinguish between **independent** and **dependent** variables. If we think something is the cause of something else, we say that *the cause is the independent variable* and that *the consequence (or the thing that is being caused) is the dependent variable*. To help you remember the difference, think that variables being caused are dependent on the causal variable, whereas causal variables are not dependent, but independent. That is why the scatterplot screen gives you the option of identifying one variable as dependent and the other as independent.

So far we have demonstrated that the social environment influences the very personal decisions about marriage and family life—but this does not mean that individual behavior is totally controlled by social forces. Even in the central states where the size of the married population is proportionately very large,

---

[1] As you'll find in the next chapter, statistical significance in survey data helps us to assess whether or not a relationship exists in the population from which the sample was drawn. In ecological data sets, such as the 50 states, statistical significance helps us determine whether the existing relationship is a result of chance factors.

there are individuals who choose not to be married. Individuals who live in very similar communities may make very different decisions about their marriage and family life. And, the choices that individuals make can also influence the social structure. Think of all of the businesses in urban areas that are developed and prosper because they serve the needs of single adults. Or, imagine how different rural areas might look if they had a larger population of singles. Schools, businesses, churches, the local government—all would be affected by a decline in the size of the married population.

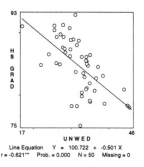

Data File: **STATES**
Task: **Scatterplot**
➤ Dependent Variable: **55) HS GRAD**
➤ Independent Variable: **29) UNWED**
➤ View: **Reg. Line**

At the bottom of the screen we see that the correlation is –0.621. This is a negative correlation. We can tell that in two ways. First, there is a minus sign preceding the correlation coefficient. Second, the regression line slopes downward from left to right, showing that as the percentage of births to unmarried women rises, the high school graduation rate decreases. The positive or negative sign signifies only the direction of the relationship—whether the two variables move in the same direction or in the opposite direction. In other words, a negative correlation of 0.621 is just as strong as a positive correlation of 0.621.

What this scatterplot tells us is that states with a higher proportion of babies born to unmarried women tend to have lower high school graduation rates. However, this does not tell us which came first—the births to unmarried women or the lower high school graduation rates. Do states have lower graduation rates because they have more women who have dropped out of school to care for a baby, or do some states have more births to unmarried women because they have high dropout rates and fewer educational opportunities? Of course, it could also be that births to unmarried women and the high school dropout rate are simply facets of another underlying cause—that both are facets of life in poverty-stricken areas.

Let's take a look at the relationships between the birth rate to unmarried women, the high school graduation rate, and the poverty rate all in one convenient table.

Data File: **STATES**
➤ Task: **Correlation**
➤ Variables: **29) UNWED**
40) %POOR
55) HS GRAD

Correlation Coefficients
PAIRWISE deletion (1-tailed test)   Significance Levels: ** = .01, * = .05

	UNWED	%POOR	HS GRAD
UNWED	1.000	0.567 **	-0.621 **
	(50)	(50)	(50)
%POOR	0.567 **	1.000	-0.632 **
	(50)	(50)	(50)
HS GRAD	-0.621 **	-0.632 **	1.000
	(50)	(50)	(50)

To construct this table, select the CORRELATION task. Select variables 29) UNWED, 40) %POOR, and 55) HS GRAD.

ExplorIt's CORRELATION task allows you to look at more than one correlation at a time. You can discover the correlation between any two variables in either of two ways. First, find one of the two variables in the horizontal list across the top of the screen. Then look down that column until you come to the second variable, as shown in the vertical list to the left of the screen. Then, you will see that the correlation between the percentage of births to unmarried women and people living below the poverty line is 0.567. You can do it the other way, too, looking for one variable at the left and then looking for the other across the top. When a variable is correlated with itself, the result is always a perfect 1.000. All correlations that meet the 0.05 level of statistical significance are indicated by one asterisk; two asterisks indicates significance above the 0.01 level. Looking at this table we see that the percentage of children born to unmarried women is associated with higher poverty rates and lower high school graduation rates and that the poverty rate and the high school graduation rate are also negatively correlated with each other.

Using these correlations, it is difficult to say whether the single-parent rate is the *cause* or the *effect* of the other two variables. Social researchers themselves differ on this point. Some believe that social problems such as poverty or dropping out of school result from the breakdown of the family. They would say that having children without being married makes it more difficult to pursue educational and career opportunities. Others believe that the breakdown of the family is actually a consequence of broader social problems. From this point of view, unmarried women may choose to have children because they don't have the opportunity for a good education and a meaningful career. Whatever the explanation, it is clear that the marriage rate to unmarried women, the poverty rate, and the high school graduation rate all appear to be related in some way. Social researchers cannot consider one of these trends without looking at the others.

Maps and correlations are useful for examining and explaining regional variations in family structures—but social scientists also look for changes in variables from one point in time to another. For example, is the percentage of married individuals in the United States the same today as it was in the 1970s? To answer that question, we can look at the results of a national survey, taken approximately every year, in which people were asked to identify their current marital status.

➤ *Data File:* **TRENDS**
   ➤ *Task:* **Historical Trends**
➤ *Variable:* **7) MARRIED%**

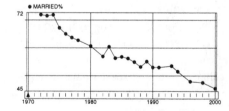

Percentage of GSS respondents who are married

Open the TRENDS data file and select the HISTORICAL TRENDS task. Select 7) MARRIED% as your trend variable.

Here we see the results for the years ranging from 1972 to 2000. In 1972 approximately 72 percent of those surveyed were married. By 2000 the percentage of married respondents drops to about 45 percent. That is a rather remarkable change for such a short period of time. Although we have freedom of choice when it comes time to decide if and when to marry, clearly forces that transcend the individual are at work as we make those decisions. The culture and social structure of the time affects the decisions we make. Living in today's culture, you are less likely to be married than if you were living in the culture of your parents when they were your same age.

What caused this downturn in the married-couples rate beginning in the mid-1970s? As we know from our analyses using the MAPPING task, the couples rate can be influenced by the size of the singles population. So, let's look at the trends for singleness for that same 1972–2000 period.

Data File: **TRENDS**
Task: **Historical Trends**
➤ Variable: **6) SINGLE%**

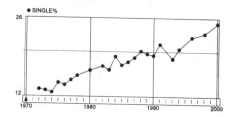

Percentage of GSS respondents who are single

This graph illustrates the percentage of single respondents who have never been married from 1972 to 2000. As you can see, the size of the singles population has risen fairly steadily during this period of time. In 1972, 13 percent of the respondents are single compared to about 25 percent in 2000. Why has the percentage of singles grown so dramatically? Part of the explanation can be found by looking at trends in education during the same period of time.

Data File: **TRENDS**
Task: **Historical Trends**
➤ Variables: **6) SINGLE%**
25) %COLLEGE

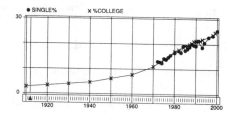

Percentage of GSS respondents who are single
Percentage of adults with college education

This graph plots both the percentage of singles and the percentage of college graduates. The data on college graduates starts with 1910 and goes to 2000, whereas the data on singles ranges from 1972 to 2000. But, if you look at the years where both data are available, you will see that the two trends closely parallel each other. The size of the singles population has risen at the same rate as the percentage of college graduates. It could be that people are delaying marriage to go to college. Or it could be that college graduates are more likely to forego marriage altogether to pursue a career. Either way, one reason for the decline in the percentage of married couples is the increase in the percentage of adults who have never married.

Do we have freedom of choice when it comes to making decisions about our marriage and family life? Yes and no. The variation in family structures we have looked at so far reveals that not everybody chooses to be married and not all families look the same. But the clear geographical, social, and chronological patterns that exist make it clear that our choices are influenced by the larger social environment. The social structure and the prevailing cultural norms of the time expand and limit the choices we have to choose from and make some options more attractive than others. But the cumulation of choices made by individuals over time also has the power to transform society and reshape the social landscape.

In the worksheet section that follows, you will have the opportunity to look at some more variations in family structures and explore how they are related to some social trends. Feel free to decide for yourself which you believe is the cause and which you think is the effect of each trend.

WORKSHEET

CHAPTER

1

NAME:

COURSE:

DATE:

REVIEW QUESTIONS

Based on the first part of this chapter, answer True or False to the following items:

The states with the highest percentage of married couples are clustered in the central portion of the United States.　　T　**F**

The states that have the highest percentage of single males also tend to have the highest percentage of single females.　　**T**　F

States with more urban residents have proportionally fewer singles.　　T　**F**

A *negative correlation* means that two variables are unrelated.　　T　**F**

Scientists accept a correlation of .500 or more as absolute proof of a causal relationship.　　T　**F**

The increase in the proportion of singles is similar to the increase in the percentage of college graduates.　　**T**　F

EXPLORIT QUESTIONS

You will need to use the ExplorIt software for the remainder of the questions. Make sure you have already gone through the "Getting Started" section that is located prior to the first chapter. If you have any difficulties using the software to obtain the appropriate information, or if you want to learn additional features of the MAPPING task, check the online help [F1].

1. Let's look at a family structure that we have not examined previously—families with children that are headed by a male with no spouse present.

> ➤ *Data File:*　**STATES**
> ➤ *Task:*　**Mapping**
> ➤ *Variable 1:*　**26) M HEAD/CH**
> ➤ *View:*　**Map**

To create this map using ExplorIt, open the STATES data file, select the MAPPING task, and select 26) M HEAD/CH as Variable 1.

a. In general, which region of the United States has the highest rate of male-headed households? (circle one)

Northeast

Midwest

Southeast

(West)

b. In general, which of these regions has the lowest rate of male-headed households? (circle one)

Midwest

(Southeast)

West

2. Now let's look at the rankings of the states by the percentage of male-headed households.

> Data File: **STATES**
> Task: **Mapping**
> Variable 1: **26) M HEAD/CH**
> ➤ View: **List: Rank**

Since you have already selected the appropriate data file, task, and variable, you only need to select the view, [List: Rank].

a. List the three states with the highest male-headed household rates, and list the percentage of male-headed households.

	STATE	% MALE-HEADED HOUSEHOLDS
1.	ALASKA	3.6
2.	New Mexico	3.1
3.	Nevada	2.9

b. List the three states with the lowest male-headed household rates, and list the percentage of male-headed households.

	STATE	% MALE-HEADED HOUSEHOLDS
48.	Alabama	1.7
49.	Connecticut	1.6
50.	Massachusetts	1.5

3. Now create a second map showing the distribution of single-parent households that are headed by a female.

> *Data File:* **STATES**
> *Task:* **Mapping**
> ➤ *Variable 1:* **25) F HEAD/CH**
> ➤ *View:* **Map**
>
> **Note that you only need to select a new variable, 25) F HEAD/CH, and the appropriate view, Map.**

   a. In general, which region of the United States has the highest rate of female-headed households? (circle one)

                                                 Northeast

                                                 Midwest

                                              (Southeast)

                                                 West

   b. In general, which region of the United States has the lowest rate of female-headed households? (circle one)

                                                 Northeast

                                                 (Midwest)

                                                 Southeast

                                                 West

4. Now look at the rankings of the states by the percentage of female-headed households.

> *Data File:* **STATES**
> *Task:* **Mapping**
> *Variable 1:* **25) F HEAD/CH**
> ➤ *View:* **List: Rank**

   a. List the three states with the highest female-headed household rates, and list the percentage of female-headed households.

	STATE	% FEMALE-HEADED HOUSEHOLDS
1.	Mississippi	10.1
2.	Louisiana	9.8
3.	Georgia	8.6

b. List the three states with the lowest female-headed household rates, and list the percentage of female-headed households.

	STATE	% FEMALE-HEADED HOUSEHOLDS
48.	*New Hampshire*	5.7
49.	*IOWA*	5.6
50.	*North Dakota*	5.3

c. Do any states have more single-parent households headed by males than by females?

Yes (No)

d. In a few sentences, describe how the distribution of male-headed households compares with that of female-headed households.

there are strong percenteges of female-headed households when compared to the male percenteges. When compared, the male-headed households lowest rate area was the females highest rate.

5. Let's see if we can explain the differences in the geographic distribution of single-parent families headed by males versus those headed by females. We'll begin by looking at the social environment surrounding female-headed households. Why does the southeastern region of the United States tend to have a greater percentage of female-headed households? Do these states have a greater percentage of females who are divorced? Or are women who live in this region of the country just less likely to have never been married but still have children?

> Data File: **STATES**
> ➤ Task: **Scatterplot**
> ➤ Dependent Variable: **25) F HEAD/CH**
> ➤ Independent Variable: **20) %DIV.FEM**
> ➤ View: **Reg. Line**

Select the SCATTERPLOT task. Note that you need to choose two variables—25) F HEAD/CH as the dependent variable and 20) %DIV.FEM as the independent variable.

a. What is the value of r for this scatterplot?

$r = \underline{0.175}$

b. Is r statistically significant?

Yes (No)

c. Do states with more divorced females have more female-headed households?

Yes (No)

6. Now let's look at the relationship between female-headed households and the proportion of single females.

> Data File: **STATES**
> Task: **Scatterplot**
> Variable 1: **25) F HEAD/CH**
> ➤ Variable 2: **29) UNWED**
> ➤ View: **Reg. Line**

Special feature: When the scatterplot is showing, you may obtain the information on any dot by clicking on it. A little box will appear around the dot, and the values of 29) UNWED (or the X-axis variable) and of 25) F HEAD/CH (or the Y-axis variable) will be shown. The information window also shows the value of the correlation coefficient with the highlighted case removed. Ignore this new value of the correlation coefficient.

a. Identify the case that is highest on UNWED.   MISSISSIPPI

b. What is its value on UNWED (X)?   45.40

c. What is its value on F HEAD/CH (Y)?   10.1

d. What is the value of r for this scatterplot?   r = .773

e. Is r statistically significant?   (Yes) No

f. States that have a high proportion of female-headed households tend to have (circle one)

    1. a lower percentage of births to unmarried women than other states.

    2. a higher percentage of births to unmarried women than other states.

    (3.) approximately the same percentage of births to unmarried women as other states.

g. Overall, it would appear that most female-headed-households probably consist of (circle one)

    1. divorced females and their children.

    (2.) single females and their children.

    3. approximately the same percentage of divorced and single females and their children.

7. Now let's make similar comparisons for single-parent families headed by a male.

> Data File: **STATES**
> Task: **Scatterplot**
> ➤ Dependent Variable: **26) M HEAD/CH**
> ➤ Independent Variable: **17) %DIV.MEN**
> ➤ View: **Reg. Line**

a. What is the value of r?   r = .556

b. Is the relationship statistically significant?   (Yes) No

c. States that have a high proportion of male-headed households tend to have (circle one)
    1. a lower percentage of divorced males than other states.
    2. a higher percentage of divorced males than other states.
    3. approximately the same percentage of divorced males as other states.

8. Now look at the relationship between male-headed households and the percentage of unmarried males.

> Data File: **STATES**
> Task: **Scatterplot**
> Dependent Variable: **26) M HEAD/CH**
> ➤ Independent Variable: **15) %SINGLE M**
> ➤ View: **Reg. Line**

a. What is the value of r?

r = 0.078

b. Is the relationship statistically significant?

Yes   (No)

c. States that have a high proportion of male-headed households tend to have (circle one)
    (1.) a lower percentage of males who have never been married than other states.
    2. a higher percentage of males who have never been married than other states.
    3. approximately the same percentage of males who have never been married as other states.

d. Overall, it would appear that most male-headed-households probably consist of (circle one)
    (1.) divorced males and their children.
    2. single males and their children.
    3. approximately the same percentage of divorced and single males and their children.

9. Now let's look at the relationship between single-parent families and poverty. Because single-parent families have only one wage-earner, do states that have higher percentages of single-parent house-holds also have higher poverty rates? Does it matter whether the single-parent families are headed by a male or a female?

> Data File: **STATES**
> ➤ Task: **Correlation**
> ➤ Variables: **25) F HEAD/CH**
>   **26) M HEAD/CH**
>   **40) %POOR**

a. What is the value of r for the relationship between F HEAD/CH and %POOR?

r = .450

b. Is it statistically significant?

(Yes)   No

   c. What is the value of r for the relationship between M HEAD/CH and %POOR?   r = *.146*.

   d. Is it statistically significant?   Yes (No)

   e. Based on these results, which of the following statements is the most accurate? (circle one)

        1. States that have more single-parent families have higher poverty rates.

        (2.) States that have more female-headed households have higher poverty rates.

        3. States that have more male-headed households have higher poverty rates.

10. Based on your intuition, the maps of marriage rates (marriages per 1,000 population) and divorce rates (divorces per 1,000 population) are likely to be (circle one of the following)

        (a.) similar to each other.

        b.   opposite of each other.

        c.   relatively unrelated to each other.

   Let's see how these maps compare.

       ➤ *Data File:* **STATES**
          ➤ *Task:* **Mapping**
       ➤ *Variable 1:* **12) MARRIAGE**
       ➤ *Variable 2:* **13) DIVORCE**
          ➤ *View:* **Map**

       The value for the correlation coefficient, Pearson's r, is indicated on the right side of the screen in the box for the lower map.

   a. What is the value of r?   r = *0.409*

   b. Is the relationship statistically significant?   (Yes) No

   c. States that have higher marriage rates tend to have (circle one)

        (1.) higher divorce rates.

        2.   lower divorce rates.

        (3.) the same divorce rate as other states.

   d. How would you explain these findings?

       Many of the states with the highest marriage rates had the same divorce rate.

11. Now let's look at marriage rates from a historical perspective.

> ➤ *Data File:* **TRENDS**
>> ➤ *Task:* **Historical Trends**
> ➤ *Variable:* **4) MARR. RATE**

Open the **TRENDS** data file and select the HISTORICAL TRENDS task. Select **4) MARR. RATE** as your trend variable.

a. In which year were marriage rates the lowest?   *1998*

b. Scroll through some of the historical events of the 1920s and 1930s. How would you explain the decline in the marriage rate that occurred during this period of time?

the Great Depression
stock market crash
~~St Valentine Mass~~

c. In which year were marriage rates the highest?   *1946*

d. Scroll through the timeline of events surrounding the period with the highest marriage rates. How would you explain this peak in marriage rates?

Due to technology increasing which brought in more money.

12. Now you can select your own maps to compare. Choose a variable you would like to map and type its name or number at the point where you are asked for Variable 1. (Remember, you can also scroll through the list of variables and double-click on the variable you wish to select.) Enter the name of the variable which you wish to use for comparison at the point where you are asked for Variable 2.

> ➤ *Data File:* **STATES**
>> ➤ *Task:* **Mapping**
> ➤ *Variable 1:* [Enter your first variable here]
> ➤ *Variable 2:* [Enter your second variable here]
>> ➤ *View:* **Map**

a. Which variables did you select?   Variable 1: *1. Females*

                                            Variable 2: *1. Fem Wrking*

Click on the [Print] button to print this map. (**Note:** If your computer is not connected to a printer or if you have been instructed not to use the printer, skip these printing instructions.) It may take some time to print a graphic such as this. Your name and the date will be printed along with the map. Attach a copy of the map to this worksheet.

b. Are the maps of the variables you selected similar to each other?   Yes   No

c. Were the results of these comparisons what you expected when you selected these variables? Explain.

the results were not what I had expected,
I expected NY.'s ratio of female workers
to be much higher. I was very surprised.

CHAPTER **2**

THE FAMILY, CULTURE, AND SOCIETY

Tasks: Mapping, Univariate, Cross-tabulation
Data Files: CULTURES, GSS

People everywhere are to some extent a product of their culture. We tend to believe that the values and norms we grew up with are natural and that any deviations from those standards are strange—or even abhorrent. For example, think of the millions of Americans who eat eggs for breakfast or have a hamburger for lunch without ever stopping to think about what they are eating. But what would happen if a restaurant were to serve lizard eggs rather than chicken eggs or horse meat rather than beef? Although these foods are common in other cultures, you may already be feeling a tightening in your stomach just thinking about eating such fare. You may not be able to articulate why one source of food is more natural than another, but the feeling is real nonetheless. In this chapter we will explore how the culture we live in and other socioeconomic characteristics similarly influence our marital and familial expectations and practices.

What are your goals when you think about marriage and family life? The prevailing norm in our society is that you would expect to fall in love, get married, and move away from home to start a life on your own. In fact, that may seem to be the *natural* order of events. But have such expectations always been held by every society, or is this another example of cultural conditioning? To answer that question, we need to look back over time to the marriage and family patterns of other cultures.

The data file we will use is based on 186 preindustrial societies from around the world. This well-known "sample" of societies was put together by George Peter Murdock and Douglas R. White and is known as the Standard Cross-Cultural Sample. It's important to understand that this data file is based on societies (not countries) and that all data are based on preindustrial times (that is, before the society became influenced by industrialization and 19th- and 20th-century modernity). Although the cultures included in this sample come from different periods in human history (the ancient Hebrews are one of the cases, for example), the data on each of the cases are from the same point in time. That is, all data for the ancient Hebrews date from Old Testament times and all the data for the Imperial Romans come from about 100B.C. In contrast, the data for the Aleut of North America date from about 1840 and those for the Pawnee Indians apply to the late 1860s.

Let's start by seeing how common it is for couples to fall in love and choose who they will marry on their own.

➤ Data File: **CULTURES**
  ➤ Task: **Mapping**
  ➤ Variable 1: **12) COUPLE PIC**
    ➤ View: **Map**
    ➤ Display: **Legend**

COUPLE PIC -- MUST BOTH BRIDE AND GROOM CONSENT TO MARRIAGE? (MGH)

Each colored circle on this map represents a society, and the legend tells you what the color of the circle represents. In this case, the dark color indicates those societies in which both the bride and the groom must consent to marriage; societies having a light-colored circle do not require consent of both the bride and the groom. The numbers on the right side of the legend are especially useful with this data file. As you can see, 26 societies (or about 35 percent) don't require the consent of both the bride and groom and 48 societies (or 65 percent) do require consent. There are no data available for 112 of the societies in this data file. Based on these results, it seems likely that most of the people living in about one-third of these preindustrial societies would have considered it rather strange to leave the decision of whom to marry up to the young people involved.

How about the American norm of married couples moving to their own private place of residence once they get married—was that the practice of other cultures?

Data File: **CULTURES**
  Task: **Mapping**
➤ Variable 1: **8) EXT.FAMILY**
  ➤ View: **Map**
  ➤ Display: **Legend**

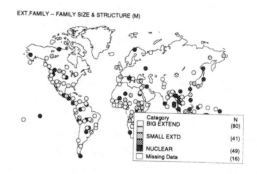

EXT.FAMILY -- FAMILY SIZE & STRUCTURE (M)

The *nuclear family* (which is the family comprised only of the wife, the husband, and their children) was clearly not the most common in preindustrial societies. Nearly half of these preindustrial societies have predominantly large *extended families*, which consist of family members beyond the wife, the husband and their children, and nearly a quarter more have small extended families. The nuclear family is the most common family form in just 49 of the 170 preindustrial societies.

If we look further into this data set, we can find other variations from family norms that most modern Americans take for granted. For example, let's look at the practice of married couples sleeping together.

Data File: **CULTURES**
Task: **Mapping**
➤ Variable 1: **19) COUPLE BED**
➤ View: **Map**
➤ Display: **Legend**

COUPLE BED -- DO HUSBANDS AND WIVES SLEEP TOGETHER IN SAME "BED?" (B&G)

Category	N
SAME BED	(43)
APART	(67)
Missing Data	(76)

As you can see, in only 43 societies (or 39 percent) do husbands and wives typically share a bed. In the remaining 67 societies (61 percent), married couples sleep apart. Again, it is likely that the people in these societies considered this to be the natural way things should be.

So far we have seen that some modern ideals of marriage and family life were not the prevailing norm in most preindustrial societies. Most preindustrial societies had different standards when it came to choosing a mate and establishing a household. But what happens if we compare only people who are living in the contemporary United States? Are there differences among different sociocultural groups in our own society?

Before looking for variations among groups, let's look at the marital status of Americans as a whole. What percentage of Americans are currently married? What percentage are divorced, separated, or widowed? What percentage have never been married?

➤ Data File: **GSS**
➤ Task: **Univariate**
➤ Primary Variable: **1) MARITAL**
➤ View: **Pie**

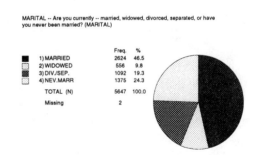

MARITAL -- Are you currently -- married, widowed, divorced, separated, or have you never been married? (MARITAL)

	Freq.	%
1) MARRIED	2624	46.5
2) WIDOWED	556	9.8
3) DIV./SEP.	1092	19.3
4) NEV.MARR	1375	24.3
TOTAL (N)	5647	100.0
Missing	2	

Here are the results of a national survey of 5,649 American adults, each of whom was interviewed at length in 1998 or 2000 as part of the General Social Survey (GSS). As you can see, 2,624 persons, or 46.5 percent of the sample, are currently married. Others were previously married but are now either widowed (9.8 percent) or divorced or separated (19.3 percent). Thus, most adults are married or have been married at some time. Note, however, that the second most frequent status in the United States is that of having never been married (24.3 percent).

The diversity of marital statuses in the above graph is evidence of the freedom of choice that Americans have when it comes to personal relationships. In contrast to many of the preindustrial societies we looked at earlier, the modern American culture tends to leave decisions about getting married, or staying married, up to individuals. Thus, we end up with a greater array of family structures. Let's explore the diversity in this sample a little further.

The GSS asked survey respondents who are not currently married whether they have a steady romantic partner or lover.

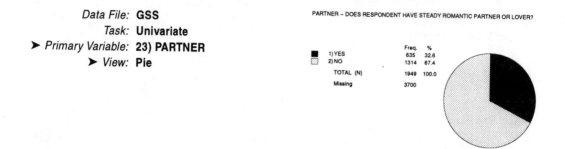

Data File: **GSS**
Task: **Univariate**
➤ Primary Variable: **23) PARTNER**
➤ View: **Pie**

PARTNER -- DOES RESPONDENT HAVE STEADY ROMANTIC PARTNER OR LOVER?

		Freq.	%
■	1) YES	635	32.6
▦	2) NO	1314	67.4
	TOTAL (N)	1949	100.0
	Missing	3700	

About 33 percent of all respondents who are not currently married indicate that they are involved in a steady, romantic relationship. How many of these respondents who have a steady relationship are living together?

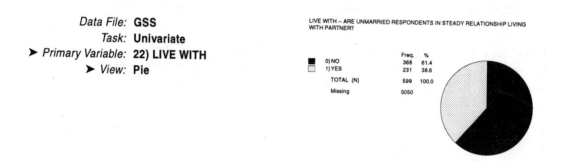

Data File: **GSS**
Task: **Univariate**
➤ Primary Variable: **22) LIVE WITH**
➤ View: **Pie**

LIVE WITH -- ARE UNMARRIED RESPONDENTS IN STEADY RELATIONSHIP LIVING WITH PARTNER?

		Freq.	%
■	0) NO	368	61.4
▦	1) YES	231	38.6
	TOTAL (N)	599	100.0
	Missing	5050	

Here we see that 38.6 percent of those who are involved in a steady relationship are cohabitating with their partner while 61.4 percent are not. Remember, this does not mean that over one-third of all singles are cohabitating because this question was asked only of people who are in a steady relationship.

Let's look at the marital status of parents in this survey.

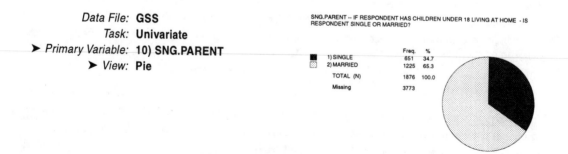

Data File: **GSS**
Task: **Univariate**
➤ Primary Variable: **10) SNG.PARENT**
➤ View: **Pie**

SNG.PARENT -- IF RESPONDENT HAS CHILDREN UNDER 18 LIVING AT HOME - IS RESPONDENT SINGLE OR MARRIED?

		Freq.	%
■	1) SINGLE	651	34.7
▦	2) MARRIED	1225	65.3
	TOTAL (N)	1876	100.0
	Missing	3773	

This graph represents the percentage of respondents who have children under the age of 18 living at home who are married or single. About one-third of the parents in this sample are unmarried (34.7 percent) and two-thirds are married (65.3 percent). Of course, some of the single parents may have been married at some time, but then became either divorced or widowed. Let's see what percentage of the single parents have never been married.

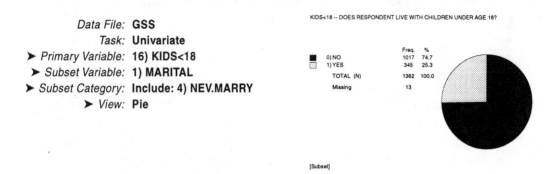

KIDS<18 -- DOES RESPONDENT LIVE WITH CHILDREN UNDER AGE 18?

		Freq.	%
■	0) NO	1017	74.7
▨	1) YES	345	25.3
	TOTAL (N)	1362	100.0
	Missing	13	

Data File: **GSS**
Task: **Univariate**
➤ Primary Variable: **16) KIDS<18**
➤ Subset Variable: **1) MARITAL**
➤ Subset Category: **Include: 4) NEV.MARRY**
➤ View: **Pie**

[Subset]

The option for selecting a subset variable is located on the same screen you use to select other variables. For this example, select 16) KIDS<18 as the Primary variable and select 1) MARITAL as a subset variable. A window will appear that shows you the categories of the subset variable. Select 4) NEV.MARRY as your subset category and choose the [Include] option. Then click [OK] and continue as usual.

With this particular subset selected, the results will be limited to those in the sample who have never been married. The subset selection continues until you exit the task, delete all subset variables, or clear all variables.

By using the subset feature, we are able to limit this graph only to those who have never been married. Here we see that about one-fourth, 25.3 percent, of the respondents who have never been married have at least one child.

So far we have seen that there is a variety of family patterns represented in this national survey. Just under half of those surveyed are currently married, about one-fourth have never married, and the rest are either widowed or divorced. Among those who are not married, about half are in a romantic relationship and about half are not. Among the half that have a steady partner, just over one-third are cohabitating. Finally, about two-thirds of all unmarried women, and one-third of never-married women, are single parents. As opposed to cultures where personal choice is limited by stringent social controls, the value Americans place on individual freedom has resulted in a wide variety of marriage and family lifestyles.

With the freedom to make choices comes the reality of having to live with the consequences of the decisions we make, thus it is important that we make informed decisions whenever possible. Marriage and family research can be a tool for understanding both the factors that influence our decisions and the likely outcome of the decisions we make. Throughout these first two chapters, we have seen evidence of how family-related decisions are influenced by the surrounding culture. Becoming aware of these social influences can help us determine whether the decision we feel inclined to make is really the *best* option or just the easiest choice to make to fit in with the surrounding culture.

Another value of social research is that it can help us understand the nature of personal relationships. For example, in the marriage and family classes that I teach, I emphasize that most married couples go through a decline in their marital happiness during their childrearing years. I say this not to be a doom-sayer, but because I want them to be prepared to respond appropriately when such a decline occurs.

Finally, marriage and family research can give us an idea as to what the consequences, positive or negative, of our decisions may be for ourselves and others. Social research cannot predict the outcome for every individual decision we make, but it can identify trends that have occurred among other people in similar circumstances. For example, let's use the CROSS-TABULATION task to look for the relationship between single parenting and family income.

Data File: **GSS**
➤ Task: **Cross-tabulation**
➤ Row Variable: **28) INCOME**
➤ Column Variable: **10) SNG.PARENT**
➤ View: **Tables**

INCOME   by   SNG.PARENT
Cramer's V: 0.506 **

		SINGLE	MARRIED	Missing	TOTAL
	LOW	279	96	986	375
	MIDDLE	240	434	1315	674
	HIGH	68	594	947	662
	Missing	64	101	525	690
	TOTAL	587	1124	3773	1711

(SNG.PARENT across top; INCOME down side)

To construct this table, return to the main menu and select the CROSS-TABULATION task. Then select 28) INCOME as the row variable and 10) SNG.PARENT as the column variable.

Across the top is the variable separating single respondents with children from married respondents with children. Down the side are the labels representing different levels of family income. The numbers within the table reflect the numbers of unmarried female respondents who gave each answer on the income question.

Looking at the first column in the table we see that of the 587 unmarried respondents with children, 279 are in the low income category, 240 are in the middle income category, and 68 are in the highest income category. Now compare the difference between parents who are single with those who are married. Of those parents who are married, 96 are in the lowest income level, 434 have middle level incomes, and 594 are in the highest income category. Clearly there are more married persons in the highest income categories. However, it is hard to say how great this difference really is since there are half as many single parents (587 total respondents) as there are married respondents (1,124 total respondents). Thus, we can't simply compare raw numbers; we must take differences in the size of the populations into account. To do so, we can calculate the **percentage** of parents in each category.

Data File: **GSS**
Task: **Cross-tabulation**
Row Variable: **28) INCOME**
Column Variable: **10) SNG.PARENT**
View: **Tables**
➤ Display: **Column %**

INCOME   by   SNG.PARENT
Cramer's V: 0.506 **

		SINGLE	MARRIED	Missing	TOTAL
	LOW	279	96	986	375
		47.5%	8.5%		21.9%
	MIDDLE	240	434	1315	674
		40.9%	38.6%		39.4%
	HIGH	68	594	947	662
		11.6%	52.8%		38.7%
	Missing	64	101	525	690
	TOTAL	587	1124	3773	1711
		100.0%	100.0%		

If continuing from the previous example, simply select the [Column %] option to view the column percentages.

Marriage and Family

Now we can more easily see the connection between single parenting and family income. Across the top row we see that among parents who are not married, 47.5 percent have low incomes, as compared to just 8.5 percent of those who are married. Looking across the bottom row we see that 11.6 percent of the single parents are in the high income category, as compared to 52.8 percent of the married parents. Of course, we do not know which came first, becoming a single parent or having a low income. It could be that unmarried adults with children are more likely to have their education interrupted and may also have less time for paid employment because of childrearing demands. In this case, being a single parent would be the shaping action—or the independent variable. On the other hand, those with lower incomes may be more likely to have a child as a source of personal fulfillment or they may be more likely to be divorced as the result of financial strains placed on their marriage. If this is true, it is income that is the independent variable.

There are social scientists who adhere to each of these perspectives. The data available for these analyses do not allow us to determine time sequences, so you will have to choose the approach you think is most plausible. But clearly there is a relationship between family structure and socioeconomic status.

Using social research to explore patterns, such as the relationship between single-parent households and income, does not allow us to make absolute predictions for individuals. Not all single parents are in the lowest income category, nor are all married parents in the highest income category. But social research can help as we consider the probable outcomes of different courses of action. And social research is also a valuable tool for those who must consider the effect that social policies are likely to have on families.

We have seen how income is related to the status of being a single parent. Now let's see how some other socioeconomic factors, such as race and income, affect marital statuses in general.

Data File: **GSS**
Task: **Cross-tabulation**
➤ Row Variable: **2) MARRIED?**
➤ Column Variable: **26) RACE**
➤ View: **Tables**
➤ Display: **Column %**

MARRIED?   by   RACE
Cramer's V: 0.151 **

		RACE			
		WHITE	BLACK	Missing	TOTAL
MARRIED?	YES	2241	244	139	2485
		50.0%	29.4%		46.8%
	NO	2238	587	198	2825
		50.0%	70.6%		53.2%
	Missing	0	1	1	2
	TOTAL	4479	831	338	5310
		100.0%	100.0%		

Here we see that there is a relationship between race and current marital status (Note: A complete study of the relationship between race and family should include Hispanics and other racial-ethnic groups; however, these groups are not represented in large enough numbers in this sample to be included in our analyses). Exactly half of the white respondents in this sample are currently married, compared to 29.4 percent of all African Americans. Of course, the unmarried category in this table includes those who are divorced or widowed along with those who have never married. So, is race also related to the percentage of respondents who have never married?

Data File: **GSS**

Task: **Cross-tabulation**

➤ Row Variable: **3) EVER MAR?**

➤ Column Variable: **26) RACE**

➤ View: **Tables**

➤ Display: **Column %**

EVER MAR?   by   RACE
Cramer's V: 0.115 **

		RACE			
		WHITE	BLACK	Missing	TOTAL
EVER MAR?	YES	3521	542	209	4063
		78.6%	65.2%		76.5%
	NO	958	289	128	1247
		21.4%	34.8%		23.5%
	Missing	0	1	1	2
	TOTAL	4479	831	338	5310
		100.0%	100.0%		

The results in this table are a little closer than the results in the previous one: 78.6 of the whites have been married at some time compared to 65.2 percent of the African Americans. Although there appears to be a difference between the two racial categories, the majority of both blacks and whites have been married at some time, so how can we say there is a difference between these two groups?

Random sampling is the basis of all survey research. Rather than interview all the members of a population, survey researchers interview only a sample. As long as this sample is selected randomly, so that all members have an equal chance of being selected, the results based on the sample can be generalized to the entire population. That is, the laws of probability allow us to **calculate the odds** that something observed in the sample accurately reflects a feature of the population sampled. These odds are determined by two factors: the size of the sample and the size of observed differences within the sample.

First of all, the sample must be **sufficiently large**. Obviously, we couldn't use a sample of two people as the basis for describing the American population: there is a very high probability that they both would be married. For this reason, survey studies include enough cases so that they can accurately reflect the population in terms of variations in such characteristics as age, sex, education, race, religion, and so on. The accuracy of a sample is a function of its size: the larger the sample, the more accurate it is. Good survey studies are usually based on 1,000 cases or more. This sample is based on 5,310 Americans, although many of the questions were asked only of 1,500 or so respondents.

The second limitation has to do with the **magnitude of the difference** observed in a table. The larger the difference we observe in a sample, the greater the probability that it is representative of the population. Because samples are based on the principle of random selection, they are subject to some degree of fluctuation. That is, for purely random reasons, there can be small differences between the sample and the population. For example, if we flipped 1,000 coins, you know that we "should" get 500 heads, but you also know that someone could easily get 482 heads and someone else 510 heads. By the same token, a sample can sometimes differ from the population for random reasons. Thus, whenever we examine cross-tabulations such as those shown above, social scientists must always ask whether they are seeing a real difference, one that would turn up if the entire population were examined, or only a random fluctuation, which does not reflect a true difference in the population.

The small size of the differences between the value placed on marriage by those who have been divorced and those who have not been divorced would make any experienced analyst suspect that the differences observed are merely the result of random fluctuations.

Fortunately, there is a very simple technique for calculating the odds that a given difference is real or random. This calculation is called a **test of statistical significance**. You will (hopefully) recall that we dealt with statistical significance in Chapter 1 when working with data based on the 50 states. The test of statistical significance for survey data works on the same principle. Differences observed in survey samples are said to be statistically significant when the odds against the results being random are high enough. Through the years, social scientists have settled on the rule of thumb that they will ignore all differences unless the odds are at least **20 to 1** against their being random. To put it another way, social scientists reject all findings when the probability that they are random is greater than .05 (5 in 100). This level of significance means that if 100 random samples were drawn independently from the same population, a statistical relationship between any two variables would not turn up more than five times, purely by chance. In fact, many social scientists think this is too lenient of a standard and some even require that the probability that a finding is random be less than .01 (1 in 100). To apply these rules of thumb, social scientists calculate the **levels of significance** of the differences in question and compare them against these standards.

There are two ways to see what the level of significance is for this table. If you want to know the exact probability of whether these results are due to random fluctuations, you need to switch to the statistics view.

Data File: **GSS**
Task: **Cross-tabulation**
Row Variable: **3) EVER MAR?**
Column Variable: **26) RACE**
➤ View: **Statistics (Summary)**

EVER MAR?   by   RACE

Nominal Statistics

Chi-Square 69.926	(DF =	1;	Prob. = 0.000)			
V:	0.115		C:	0.114		
Lambda (DV=26)	0.000		Lambda (DV=3)	0.000	Lambda	0.000

Ordinal Statistics

Gamma	0.324	Tau-b:	0.115	Tau-c:	0.071
s.error	0.043	s.error	0.015	s.error	0.009
Dyx	0.134	Dxy.	0.098		
s.error	0.018	s.error	0.013		
Prob. =	0.000				

Ignore everything on this screen except for the first two lines of text. At the end of the first line you'll see "Prob. = 0.000." This value indicates that the odds the cross-tabulation results are simply due to randomness are less than 1 in 1,000 (if the odds were 1 in 1,000, the probability would be .001, but since the probability is 0.000 that means the odds are even less than 1 in 1,000). Since social scientists require that these odds be less than 1 in 20 (.05) or even 1 in 100 (.01), we know that the differences between whites and blacks who have ever been married are probably not due to random fluctuations.

There is another number on this screen that you will find useful when doing cross-tabulation analysis. In the second row, locate the value "V = .115." The V stands for Cramer's V, which is a correlation coefficient developed for cross-tabulations. Cramer's V is similar to Pearson's r (see Chapter 1) in that its value varies from 0 to 1. If the relationship between these two variables was perfect (that is, if all whites had ever been married and all blacks had not, or vice versa), then this would be a value of 1. However, unlike Pearson's r, V does not indicate whether a relationship is positive or negative—this must be inferred from the table itself. In addition, V is much less sensitive than r. A simple rule of thumb might help you determine the strength of V. If V is between .00 and .10, the relationship is very weak or nonexistent; if it's between .10 and .25, it should be considered a moderate relationship; and any V over .25 should be considered a strong relationship.

It's important to note that Cramer's V assesses the *strength* of the relationship shown in the cross-tabulation, *not* the odds that it is statistically significant. Don't confuse Cramer's V with the probability value—they provide different information.

Here's something else to remember. Sometimes in large surveys like the GSS, you'll see percentage differences in the table of only 2 to 3 percent that are statistically significant. Even if this difference does exist in the entire population, it's generally not worthy of note. Whenever you examine a cross-tabulation in this book, always look at the actual difference in percentage points. If the difference is less than 8 or 10 percentage points, you should ask yourself whether it is *substantively significant*.

To summarize, whenever you do a cross-tabulation, you should examine four things. First, look at the table to see if the observed differences are worth noting. If the categories you are comparing differ by less than 8 or 10 percentage points, it might not be worth pursuing the analysis. Second, if you are testing a hypothesis, be sure that the percentages differ in the predicted direction. For example, if you predict that whites are more likely to be married than African Americans, then the differences in the table should support this. If you make it through the first two steps, you should further determine the strength of the relationship by examining the correlation coefficient (V). Finally, determine whether your results are statistically significant. That's when you should use the .01 and .05 rules discussed previously.

ExplorIt makes it easy to go through these steps by placing all the information you need on one screen. Return to the column percentages view for this cross-tabulation. Notice that the value of V also appears on this screen. If V is followed by *no* asterisks, the difference it represents is not statistically significant, and we conclude that there is no difference in the U.S. population. If V is followed by one asterisk, the difference it represents is statistically significant at the .05 level of probability. If V is followed by two asterisks, as it is in this case, the difference it represents is statistically significant at the .01 level of probability.

We now know that African Americans are somewhat less likely than whites to have ever been married. Let's pursue this a little further by testing some hypotheses related to this observation. Our first hypothesis is that **African Americans will be less likely than whites to say they would like to get married when they meet the right person.**

Data File: **GSS**	
Task: **Cross-tabulation**	
➤ *Row Variable:* **21) WILL WED2**	
➤ *Column Variable:* **26) RACE**	
➤ *View:* **Tables**	
➤ *Display:* **Column %**	

WILL WED2   by   RACE
Cramer's V: 0.016

		RACE			
		WHITE	BLACK	Missing	TOTAL
WILL WED2	YES	140	37	16	177
		50.2%	52.1%		50.6%
	NO	139	34	8	173
		49.8%	47.9%		49.4%
	Missing	4200	761	314	5275
	TOTAL	279	71	338	350
		100.0%	100.0%		

Right away we see that the results in this table do not meet the standards for statistical significance. First off, the percentage of whites (50.2 percent) who say they would like to get married when they meet the right person is lower than the percentage for African Americans (52.1 percent). So the percentages go in the opposite direction of what was predicted in the hypothesis. Secondly, the difference is too small to be of any substantive significance. Finally, the value of V is only .016 and is not followed

Marriage and Family

by any asterisks, so it is not statistically significant. Based on these results, our hypothesis is rejected. African Americans are no less likely than whites to say they would like to get married when they meet the right person.

If the difference in marriage rates for blacks and whites is not the result of the desirability of marriage, then it must be related to other causes. Earlier we saw that there was a relationship between single parenthood and family income. Let's see if a similar relationship exists in the entire sample between a person's own income and the likelihood that they have ever been married. Our hypothesis is that **those with high incomes are more likely to have ever been married than those with low incomes**.

Data File:	**GSS**
Task:	**Cross-tabulation**
➤ *Row Variable:*	**3) EVER MAR?**
➤ *Column Variable:*	**29) R.INCOME**
➤ *View:*	**Tables**
➤ *Display:*	**Column %**

EVER MAR?   by   R.INCOME
Cramer's V: 0.151 **

		R.INCOME				
		$0K-17.4K	17.5K-34.9	$35K +	Missing	TOTAL
EVER MAR?	YES	783	938	927	1624	2648
		63.9%	72.6%	80.7%		72.3%
	NO	442	354	221	358	1017
		36.1%	27.4%	19.3%		27.7%
	Missing	2	0	0	0	2
	TOTAL	1225	1292	1148	1982	3665
		100.0%	100.0%	100.0%		

Our hypothesis appears to be supported. Among those in the lowest income category, 63.9 percent have been married; the percentage rises to 72.6 percent in the middle income category and 80.7 percent in the highest income category. The relationship is statistically significant (V = .151**); those with higher incomes are more likely to get married. Again, the direction of causality probably goes both ways. Having a low income can make it more difficult to find a desirable mate, and financial stress may strain relationships. And those who are married may be able to devote more time to their careers if they have partners with whom to share childcare and housework responsibilities.

Now that we know there is a relationship between income and marital status, we need to determine whether it is influencing the relationship between race and marital status. In other words, is the relationship between race and marital status really the result of racial differences, or is it a by-product of income differences between blacks and whites? To answer that question, we need to use the control variable option in the CROSS-TABULATION task.

Data File:	**GSS**
Task:	**Cross-tabulation**
Row Variable:	**3) EVER MAR**
➤ *Column Variable:*	**26) RACE**
➤ *Control Variable:*	**29) R.INCOME**
➤ *View:*	**Tables: $0K–$17.4K**
➤ *Display:*	**Column %**

EVER MAR?   by   RACE
Controls: R.INCOME: $0K-17.4K
Cramer's V: 0.109 **

		RACE			
		WHITE	BLACK	Missing	TOTAL
EVER MAR?	YES	602	127	54	729
		67.6%	54.7%		64.9%
	NO	289	105	48	394
		32.4%	45.3%		35.1%
	Missing	0	1	1	2
	TOTAL	891	232	103	1123
		100.0%	100.0%		

The option of selecting a control variable is located on the same screen you use to select other variables. For this example, select 29) R.INCOME as a control variable and then click [OK] to continue as usual. Separate tables for each of the 29) R.INCOME categories will now be shown for the 3) EVER MAR and 26) RACE cross-tabulation.

At the top left of the screen, you will see R.INCOME $0K–17.4K. This means that this table is limited to those who are in this lowest income category. You will see that the results are similar to those in the previous table, before the control variable was added. There is about a 13 percent difference between ever-married percentages for whites (67.6 percent) and blacks (54.7 percent). The probability is statistically significant ($V = .109^{**}$).

Now let's look at the results for those respondents in the middle income category.

Data File: **GSS**
Task: **Cross-tabulation**
Row Variable: **3) EVER MAR**
Column Variable: **26) RACE**
Control Variable: **29) R.INCOME**
➤ View: **Tables: $17.5K–$34.9K**
➤ Display: **Column %**

EVER MAR?   by   RACE
Controls: R.INCOME: 17.5K-34.9
Cramer's V: 0.013

		RACE			
		WHITE	BLACK	Missing	TOTAL
EVER MAR?	YES	764	126	48	890
		73.6%	72.0%		73.4%
	NO	274	49	31	323
		26.4%	28.0%		26.6%
	TOTAL	1038	175	79	1213
		100.0%	100.0%		

Click the ➤ button at the bottom of the task bar to look at the second (or "next") partial table for 29) R.INCOME.

In this table, which includes only those in the middle income category, there is virtually no difference between the percentage of whites (73.6 percent) and blacks (72 percent) who have ever been married. As we would expect, this slight difference is not statistically significant ($V = .013$). Let's check the last table.

Data File: **GSS**
Task: **Cross-tabulation**
Row Variable: **3) EVER MAR**
Column Variable: **26) RACE**
Control Variable: **29) R.INCOME**
➤ View: **Tables: $35K +**
➤ Display: **Column %**

EVER MAR?   by   RACE
Controls: R.INCOME: $35K +
Cramer's V: 0.046

		RACE			
		WHITE	BLACK	Missing	TOTAL
EVER MAR?	YES	803	77	47	880
		81.7%	75.5%		81.1%
	NO	180	25	16	205
		18.3%	24.5%		18.9%
	TOTAL	983	102	63	1085
		100.0%	100.0%		

The difference between the percentage of whites (81.7 percent) and blacks (75.5 percent) in the highest income category who have ever been married is a little larger than the difference in the previous table, but it is still not statistically significant ($V = .046$).

What these results tell us is that the difference in the likelihood of blacks or whites having ever been married is partially accounted for by social class differences. Race has a statistically significant impact on the likelihood of marriage in the lowest income category, but not in the two highest income categories. However, if you look back at the tables, you will see that there were more African-American respondents in the lowest income category (232 respondents) than in either the middle (175 respondents) or upper (102 respondents) income categories. This helps explain why there is a statistically significant difference between the percentage of blacks and whites who have ever been married overall—nearly half of the African-American respondents are in the lowest income category.

The likelihood of marriage does vary by race, and studies that have used larger samples of African Americans have found that this relationship cannot be totally explained away by economic inequality. But the results of our analysis demonstrate the importance of considering the influence of economic inequality when looking at the relationship between race and marriage. In the worksheet section that follows, you will have the opportunity to explore some more socioeconomic factors that influence the family.

CHAPTER

REVIEW QUESTIONS

Based on the first part of this chapter, answer True or False to the following items:

Most preindustrial societies did not require both the bride and the groom to consent to a marriage.　　　T　(F)

The nuclear family was the predominant family structure in preindustrial societies.　　　T　(F)

About one-third of the parents in the GSS are unmarried.　　　T　(F)

An observed difference is statistically significant if the probability is less than 50 percent.　　　(T)　F

Cramer's V is similar to Pearson's r in that both are correlation coefficients.　　　T　(F)

Cramer's V does not indicate whether a relationship is positive or negative. This must be inferred from the table itself.　　　(T)　F

African Americans are less likely than whites to be married.　　　(T)　(F)

People with higher incomes are less likely to have ever been married.　　　(T)　(F)

EXPLORIT QUESTIONS

If you have any difficulties using the software to obtain the appropriate information, or if you want to learn additional features of the MAPPING task, check the online help [F1].

1. In the first part of this chapter, we found that those with higher incomes are more likely to have ever been married than those with lower incomes. Because income and education are closely related, let's test the hypothesis that **the more education someone has, the more likely he or she is to be married.**

> ➤ Data File: **GSS**
> ➤ Task: **Cross-tabulation**
> ➤ Row Variable: **2) MARRIED?**
> ➤ Column Variable: **27) EDUCATION**
> ➤ View: **Tables**
> ➤ Display: **Column %**

To construct this table, select the CROSS-TABULATION task from the main menu, then select 2) MARRIED? as the row variable and 27) EDUCATION as the column variable. Select the [Column %] option to view the column percentages.

a. Use column percentages to fill in the table below.

	NOT HS GRAD	HS GRAD	COLL EDUC
YES	37.5 %	47.6 %	48.7 %

b. What is the value of V?   V = .083

c. Is V statistically significant?   (Yes)   No

d. When it comes to being married, which group do high school graduates closely resemble? (circle one)

   1. those who did not complete high school
   (2.) those who have attended college

e. Statistically speaking, which of the following statements is supported by these results? (circle one)

   1. Education does not have an impact on a person's current marital status.
   (2) Those who have not completed high school are less likely to be married.
   3. Those who have attended college are far more likely than those who have not attended college to be married.

2. According to the state-level data we examined in the first chapter of this workbook, states with more rural populations tend to have more married couples. So, let's test the hypothesis that **people who live in rural areas will be more likely to be married.**

    Data File: **GSS**
    Task: **Cross-tabulation**
    Row Variable: **2) MARRIED?**
    ➤ Column Variable: **38) COMMUNITY**
    ➤ View: **Tables**
    ➤ Display: **Column %**

a. Use column percentages to fill in the table below.

	BIG CITY	SUBURBS	SMALL TOWN	RURAL
YES	37.0 %	46.9 %	42.9 %	64.4 %

b. What is the value of V?   V = .150

c. Is V statistically significant?   (Yes)   No

d. Is the hypothesis supported? Yes   No

3. Let's take a closer look at the single-parent families.

   a. What percentage of the single-parents do you predict will be male?   _____25_____ %

   > Data File: **GSS**
   > ➤ Task: **Univariate**
   > ➤ Primary Variable: **25) SEX**
   > ➤ Subset Variable: **10) SNG.PARENT**
   > ➤ Subset Category: **Include: 1) SINGLE**
   > ➤ View: **Pie**

   > To construct this table, return to the main menu and select the UNIVARIATE task. Select 25) SEX as the primary variable and 10) SNG.PARNT as the subset variable. A window will appear that shows you the categories of the subset variable. Select 1) SINGLE as your subset category and choose the [Include] option. Then click [OK] and continue as usual.

   b. What percentage of the single parents in this sample are male?   _____21.5_____ %

   c. What percentage are female?   _____78.5_____ %

4. Single-parent families can be the result of having a child without being married, divorce, or the death of a parent. Let's see which of these patterns is the most common.

   > Data File: **GSS**
   > Task: **Univariate**
   > ➤ Primary Variable: **1) MARITAL**
   > ➤ Subset Variable: **10) SNG.PARENT**
   > ➤ Subset Category: **Include: 1) SINGLE**
   > ➤ View: **Pie**

   > Hint: If you replace only the primary variable, the subset variable will remain with your previously selected categories.

   a. What percentage of single parents are widows?   _____8.9_____ %

   b. What percentage of single parents are divorced?   _____53.9_____ %

   c. What percentage of single parents have never been married?   _____37.2_____ %

5. In the first part of this chapter, we saw that single parents tend to have lower incomes. Let's see if it makes a difference economically if the single parent is divorced or has never been married.

              Data File: **GSS**
       ➤ *Task:* **Cross-tabulation**
    ➤ *Row Variable:* **28) INCOME**
 ➤ *Column Variable:* **1) MARITAL**
  ➤ *Subset Variable:* **10) SNG.PARENT**
 ➤ *Subset Category:* **Include: 1)SINGLE**
       ➤ *View:* **Tables**
     ➤ *Display:* **Column %**

The [Subset] feature for the CROSS-TABULATION task works the same way as the [Subset] feature in the UNIVARIATE task. After you enter the subset variable, in this case 10) SNG.PARENT, a window will appear so that you can select to include 1) SINGLE.

a. Use column percentages to fill in the table below.

	WIDOWED	DIV./SEP.	NEV. MARR
LOW	~~13~~ %	~~141~~ %	_____ %
	26	43.8	58.1

b. What is the value of V?   V = .140

c. Is V statistically significant?   (Yes) No

d. Which of the following types of single parents is most likely to have the lowest income? (circle one)

    (1.) a single parent who is a widow
    2. a single parent who is divorced
    3. a single parent who has never been married

e. How would you explain these results?

After marriage and children most women do not work and concern themselves with child care and let the spouse take care of the income.

6. In the first part of this chapter, we saw that there is a relationship between race and current marital status. Is there also a relationship between race and being a single parent?

              Data File: **GSS**
          Task: **Cross-tabulation**
    ➤ *Row Variable:* **10) SNG.PARENT**
 ➤ *Column Variable:* **26) RACE**
       ➤ *View:* **Tables**
     ➤ *Display:* **Column %**

> Hint: Click [Clear All] to remove the subset variable before proceeding with this cross-tabulation.

a. What percentage of African Americans with children are single parents? ___62.7__ %

b. What percentage of whites with children are single parents? ___24.8__ %

c. What is the value of V?   V = __.300__

d. Is it statistically significant?   (Yes)   No

7. Earlier we saw that the relationship between race and having ever been married occurred primarily among those in the lowest income category. Let's see if the same trend occurs when looking at the relationship between race and single parenthood. To do this we will use the subset feature to exclude those in the lowest income category.

> Data File: **GSS**
> Task: **Cross-tabulation**
> Row Variable: **10) SNG.PARENT**
> Column Variable: **26) RACE**
> ➤ Subset Variable: **29) R.INCOME**
> ➤ Subset Category: **Exclude: 1) $0K–17.4K**
> ➤ View: **Tables**
> ➤ Display: **Column %**

> Be sure to click [Exclude] after you select the lowest income category.

a. Use column percentages to fill in the table below.

	WHITE	BLACK
SINGLE	_23.7_ %	_48.9_ %

b. What is the value of V?   V = __.205__

c. Is V statistically significant?   (Yes)   No

d. Are African Americans more likely than whites to be single parents when you control for income?   (Yes)   No

8. The previous table looked at the relationship between race and the status of being a single parent for the respondents themselves. Let's see if the pattern is similar for the marital status of the respondents' parents.

Data File: **GSS**
Task: **Cross-tabulation**
➤ Row Variable: **45) FAMILY @16**
➤ Column Variable: **26) RACE**
➤ View: **Tables**
➤ Display: **Column %**

Be sure to remove the subset variable before continuing.

a. Use column percentages to fill in the *bottom* row of the table below.

	WHITE	**BLACK**
NO	26.9 %	50.0 %

b. What is the value of V?

V = .182

c. Is V statistically significant?

(Yes)   No

d. Use the results of this table and the table in Question 6 to complete the following table.

	WHITE	**BLACK**
Percentage of respondents who are single parents		
Percentage of respondents raised by single parents		

e. The percentage of whites who are single parents themselves is (circle one)
   1. larger than the percentage who were raised by one parent.
   2. smaller than the percentage who were raised by one parent.
   3. about the same as the percentage who were raised by one parent.

f. The percentage of blacks who are single parents themselves is (circle one)
   1. larger than the percentage who were raised by one parent.
   2. smaller than the percentage who were raised by one parent.
   3. about the same as the percentage who were raised by one parent.

9. Does the age at which someone has a child affect the likelihood of him or her getting married?

   a. Do you think those who have a child before they are 20 years of age will be more or less likely to have ever married? (circle one)
      1. more likely
      2. less likely

b. Why do you think the result you predicted will occur?

[handwritten] They are more likely because to marry because they have more time to work on a relation than they will be have likely to marry at an early age because they ...

Now let's look at the results.

        Data File: **GSS**
        Task: **Cross-tabulation**
   ➤ *Row Variable:* **3) EVER MAR?**
 ➤ *Column Variable:* **18) AGE KD BRN**
       ➤ *View:* **Tables**
    ➤ *Display:* **Column %**

c. Use column percentages in the *bottom* row of the table to fill in the table below.

	<20	20–29	30 AND UP
NO	17.3 %	6.4 %	4.2 %

d. Are these results consistent with what you predicted?
                                                (Yes) No

10. Throughout this chapter you have explored some of the variations in the modern American family. Some social researchers believe the changes that have occurred in the family are signs that the family is in decline. Others believe today's families are neither worse nor better than families in the past— they are just different. What do you believe? Write a paragraph using the results of this chapter to explain why you think the modern American family is, or is not, in decline.

GENDER ROLES

Tasks: Cross-tabulation, Auto-Analyzer, Univariate, Mapping, Scatterplot, Correlation
Data Files: GSS, CULTURES, GLOBAL

Although the terms sex and gender are commonly used interchangeably, social scientists make a distinction between these two concepts. *Sex* refers to the *biological* characteristic of being female or male. *Gender* refers to *masculinity* and *femininity*—the *social* characteristics associated with being male or female. For example, most cultures have traditionally associated instrumental character traits, skills that are goal or task oriented, with the male sex. Expressive character traits, those that involve nurturing and sensitivity, have been traditionally associated with the female sex. When definitions of masculinity or femininity become associated with specific social responsibilities, social scientists refer to those as *gender roles*. Women, for instance, have been expected to assume most of the responsibility for child care because nurturing has been seen as a female trait. Men have have been expected to put their energies into protecting and providing for the family.

Because gender roles are socially defined, they change as the society changes. For example, less than a century ago, education was primarily considered a masculine pursuit and most women did not even have the option of attending college. Today, a majority of college students are female. Gender role expectations have become more flexible in other institutions as well. There are an increasing number of female leaders in politics, business, religion, and medicine. It has become more acceptable for men to emphasize the importance of familial roles and to share in the responsibility of household chores. Today, there seem to be more choices available to both women and men than ever before.

Throughout the first two chapters of this workbook, we have explored the interplay between personal choice and the cultural and social forces that influence our lives. In this chapter, we will examine how the society in which we live influences our gender-related decisions. We will begin with some cross-cultural comparisons, and then turn our attention to the diversity of gender role attitudes in the United States.

In the CULTURES data file, the variable TOUGH BOYS measures the amount of emphasis a society places on boys being aggressive and competitive.

➤ *Data File:* **CULTURES**
➤ *Task:* **Univariate**
➤ *Primary Variable:* **42) TOUGH BOYS**
➤ *View:* **Pie**

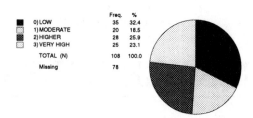

TOUGH BOYS -- IN LATE CHILDHOOD: COMBINED EMPHASIS ON FORTITUDE, AGGRESSION & COMPETITIVENESS (BOYS) (BJLM)

	Freq.	%
0) LOW	35	32.4
1) MODERATE	20	18.5
2) HIGHER	28	25.9
3) VERY HIGH	25	23.1
TOTAL (N)	108	100.0
Missing	78	

There are data for 108 societies and they are distributed fairly evenly across the various categories. Cross-tabulate TOUGH BOYS with a similar variable that's been coded for girls, TOUGH GIRL. Do societies that place an emphasis on boys being tough also encourage girls to be tough? Or do societies place a different emphasis on this trait depending on the sex of the child?

Data File: **CULTURES**
➤ *Task:* **Cross-tabulation**
➤ *Row Variable:* **44) TOUGH GIRL**
➤ *Column Variable:* **42) TOUGH BOYS**
➤ *View:* **Tables**
➤ *Display:* **Column %**

TOUGH GIRL   by   TOUGH BOYS
Cramer's V: 0.607 **
Warning: Potential significance problem. Check row and column totals.

		TOUGH BOYS					
		LOW	MODERATE	HIGHER	VERY HIGH	Missing	TOTAL
TOUGH GIRL	LOW	29	5	5	5	2	44
		96.7%	27.8%	23.8%	26.3%		50.0%
	MODERATE	1	12	5	1	0	19
		3.3%	66.7%	23.8%	5.3%		21.6%
	HIGHER	0	1	11	4	1	16
		0.0%	5.6%	52.4%	21.1%		18.2%
	VERY HIGH	0	0	0	9	0	9
		0.0%	0.0%	0.0%	47.4%		10.2%
	Missing	5	2	7	6	75	95
	TOTAL	30	18	21	19	78	88
		100.0%	100.0%	100.0%	100.0%		

Carefully examine the results in this table. Notice the downward diagonal that starts in the top left corner of the table—96.7 percent, 66.7 percent, 52.4 percent, and 47.4 percent. This indicates that the largest percentage for each category of the TOUGH BOYS variable occurs in the parallel category for the TOUGH GIRL variable. Or, to put it more simply, societies that place a low emphasis on boys being tough are most likely to place a low emphasis on girls being tough; societies that place a moderate emphasis on boys being tough are most likely to place a moderate emphasis on girls being tough; and so on. However, a closer look at this table reveals an interesting trend. As you move down the diagonal in the table, you see that as greater emphasis is placed on boys being tough, less emphasis is placed on girls being tough. In fact, less than half (47.4 percent) of the societies that place very high emphasis on boys being tough also place a high emphasis on girls being tough. In short, if a society has an unequal emphasis on toughness, it usually means that boys, rather than girls, are being encouraged to be tough.

Do males in preindustrial societies help with domestic chores?

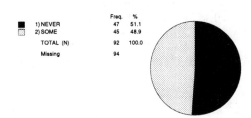

Data File: **CULTURES**
Task: **Univariate**
➤ Primary Variable: **47) HOUSEWORK?**
➤ View: **Pie**

HOUSEWORK? -- DO MALES HELP WITH DOMESTIC CHORES? (W)

		Freq.	%
■	1) NEVER	47	51.1
▨	2) SOME	45	48.9
	TOTAL (N)	92	100.0
	Missing	94	

In about half of preindustrial societies, housework is *sometimes* done by males—but that also means that it is never done by males in half of the societies.

Open the GLOBAL file so we can examine current attitudes across the world on gender roles. The World Values Survey asked people whether they think sharing household chores is important to a happy marriage. Here are the results for several dozen countries.

CHORES?: Percent who think that "sharing household chores" is "very important" to a happy marriage

➤ Data File: **GLOBAL**
➤ Task: **Mapping**
➤ Variable 1: **40) CHORES?**
➤ View: **List: Rank**

RANK	CASE NAME	VALUE
1	Nigeria	74
2	Chile	62
3	India	54
4	Argentina	53
4	Canada	53
6	Portugal	50
7	Sweden	48
7	Denmark	48
9	United States	47
9	Mexico	47

Nigerians (74 percent) are the most likely to think that sharing household chores is very important to a happy marriage. They are followed by Chileans (62 percent), Indians (54 percent), and Argentines (53 percent). The United States shows up at ninth on this list at 47 percent. At the bottom of this list we find Germany (24 percent), Estonia (19 percent), and Japan (10 percent).

These people were also asked whether they agreed that "what women really want is a home and children."

HOME&KIDS: Percent who agree that what "women really want is a home and children"

RANK	CASE NAME	VALUE
1	Lithuania	97
2	India	94
3	Slovak Republic	93
3	Czech Republic	93
5	Latvia	90
5	Bulgaria	90
5	Russia	90
8	Turkey	88
8	Nigeria	88
10	Estonia	85

Data File: **GLOBAL**
Task: **Mapping**
➤ Variable 1: **39) HOME&KIDS**
➤ View: **List: Rank**

People from countries in the former Soviet Union (e.g., Lithuania, Czech Republic, Slovak Republic, Bulgaria, Latvia, Russia) are the most likely to agree with this statement. At the bottom of this list (those less likely to agree with this statement) are highly developed Western countries (e.g., Denmark, Canada, United Kingdom, United States). Once again the range in responses is broad, with 97 percent of people in Lithuania agreeing with this statement compared to 25 percent of people from Denmark.

Let's now shift our attention to the United States and examine attitudes and actions associated with women working in the paid labor-force. We'll begin by comparing men's and women's attitudes toward female labor-force participation in general. The following GSS question was asked: "Do you approve or disapprove of a married woman earning money in business or industry if she has a husband capable of supporting her?"

➤ Data File: **GSS**
➤ Task: **Cross-tabulation**
➤ Row Variable: **67) WOMEN WORK**
➤ Column Variable: **25) SEX**
➤ View: **Tables**
➤ Display: **Column %**

WOMEN WORK by SEX
Cramer's V: 0.009

		SEX		
		MALE	FEMALE	TOTAL
WOMEN WORK	APPROVE	659	851	1510
		82.6%	81.9%	82.2%
	DISAPPROVE	139	188	327
		17.4%	18.1%	17.8%
	Missing	1663	2149	3812
	TOTAL	798	1039	1837
		100.0%	100.0%	

The vast majority of both men (82.6 percent) and women (81.9 percent) approve of married women in the workforce even if their husbands can support them. Does this mean that with regard to gender roles there is equality between the sexes, or are people just answering this question in a way that seems socially acceptable? Let's look at some more detailed questions that will help us make this distinction.

This next question asks respondents whether they agree with the statement "It is much better for everyone involved if the man is the achiever outside the home and the woman takes care of the house and family."

Marriage and Family

Data File: **GSS**

Task: **Cross-tabulation**

➤ Row Variable: **66) WIFE@HOME**

➤ Column Variable: **25) SEX**

➤ View: **Tables**

➤ Display: **Column %**

WIFE@HOME   by   SEX
Cramer's V: 0.021

		SEX		
		MALE	FEMALE	TOTAL
W I F E @ H O M E	AGREE	608	766	1374
		39.1%	37.0%	37.9%
	DISAGREE	946	1302	2248
		60.9%	63.0%	62.1%
	Missing	907	1120	2027
	TOTAL	1554	2068	3622
		100.0%	100.0%	

Less than half of men and women agree with this statement. As in the previous table, there is not a statistically significant difference between the attitudes of men (39.1 percent) and women (37 percent). Notice, however, that the respondents are more evenly divided on this question, which asked if it is best that a wife not work outside the home, than on the previous question, which asked if it is *acceptable* for women to work outside the home.

Now take this one step further. How will men and women respond when asked whether a mother who is employed can have just as loving and secure relationship with her child as a mother who is not employed?

Data File: **GSS**

Task: **Cross-tabulation**

➤ Row Variable: **64) MOTH.WORK?**

➤ Column Variable: **25) SEX**

➤ View: **Tables**

➤ Display: **Column %**

MOTH.WORK?   by   SEX
Cramer's V: 0.149 **

		SEX		
		MALE	FEMALE	TOTAL
M O T H . W O R K ?	AGREE	892	1484	2376
		56.2%	70.7%	64.5%
	DISAGREE	694	616	1310
		43.8%	29.3%	35.5%
	Missing	875	1088	1963
	TOTAL	1586	2100	3686
		100.0%	100.0%	

A majority of both sexes agree that employed mothers can have the same quality of relationships with their children as mothers who are not employed. However, women (70.7 percent) are more likely than men (56.2 percent) to agree with this item. This difference is statistically significant (V = .149**). What if the children are of preschool age—do men and women still hold the same attitudes?

Data File: **GSS**

Task: **Cross-tabulation**

➤ Row Variable: **65) PRESCH.WRK**

➤ Column Variable: **25) SEX**

➤ View: **Tables**

➤ Display: **Column %**

PRESCH.WRK   by   SEX
Cramer's V: 0.129 **

		SEX		
		MALE	FEMALE	TOTAL
P R E S C H . W R K	AGREE	810	813	1623
		52.3%	39.4%	44.9%
	DISAGREE	739	1253	1992
		47.7%	60.6%	55.1%
	Missing	912	1122	2034
	TOTAL	1549	2066	3615
		100.0%	100.0%	

In all of the previous tables, the majority of men and women were in agreement on the gender role questions. But here a little more than half (52.3 percent) of the men agree that preschool-age children suffer when their mothers are employed while less than half of the women (39.4 percent) hold this belief.

Let's look for some more sociodemographic patterns relating to gender role attitudes using ExplorIt's Auto-Analyzer. The Auto-Analyzer combines the univariate and cross-tabulation procedures you've already used. It first shows you the distribution of a primary variable you select and then allows you to choose one of nine demographic variables—sex, race, political party, marital status, religious preference, region, age, education, and income—to see what difference, if any, this demographic variable makes. It then gives you the appropriate cross-tabulation and actually tells you what is happening in the table. Let's see how it works.

Data File: **GSS**
➤ Task: **Auto-Analyzer**
➤ Variable: **66) WIFE@HOME**
➤ View: **Univariate**

WIFE@HOME -- AGREE OR DISAGREE?: It is much better for everyone involved if the man is the achiever outside the home and the woman takes care of the home and family. (FEFAM)

	%
AGREE	37.9%
DISAGREE	62.1%
Number of cases	3622

Among all respondents, 37.9% of the sample think it is much better if the woman takes care of the home.

To obtain these results, return to the main menu and select the AUTO-ANALYZER task. Then select 66) WIFE@HOME as your variable and click [OK].

Here's the univariate distribution for 66) WIFE@HOME. As we saw earlier, most people disagree with the statement "It is much better for everyone involved if the man is the achiever outside the home and the woman takes care of the house and family."

Now let's look at some of the demographic variables that may affect gender role attitudes. One of the easiest ways to observe how gender roles have changed over time is by comparing different age groups. If gender roles are changing over time, we can expect the following hypothesis to be supported: **Younger people will be less likely than older people to believe that women should take care of the home and family.**

Data File: **GSS**
Task: **Auto-Analyzer**
Variable: **66) WIFE@HOME**
➤ View: **Age**

WIFE@HOME -- AGREE OR DISAGREE?: It is much better for everyone involved if the man is the achiever outside the home and the woman takes care of the home and family. (FEFAM)

	<30	30-49	50 AND UP
AGREE	23.2%	30.2%	55.6%
DISAGREE	76.8%	69.8%	44.4%
Number of cases	665	1656	1292

The higher the age, the more likely the individual is to think it is much better if the woman takes care of the home. All differences are statistically significant.

The results are consistent with the hypothesis. The percentage of respondents who believe that women should primarily take care of the home and family increases with each age category. Respondents under the age of 30 are the least likely to agree (23.2 percent), and those age 50 and older are the most likely to agree (55.6 percent). The differences are statistically significant. The hypothesis is supported.

Education is another factor that is generally associated with changing gender role attitudes. Because college students are encouraged to critically evaluate traditional social norms, we would expect that **people who attend college will be less likely to believe that women should take care of the home and family**.

Data File: **GSS**
Task: **Auto-Analyzer**
Variable: **66) WIFE@HOME**
➤ View: **Education**

WIFE@HOME -- AGREE OR DISAGREE?: It is much better for everyone involved if the man is the achiever outside the home and the woman takes care of the home and family. (FEFAM)

	NO HS GRAD	HS GRAD	COLL EDUC
AGREE	56.3%	42.6%	29.6%
DISAGREE	43.7%	57.4%	70.4%
Number of cases	607	1057	1942

The higher the education, the less likely the individual is to think it is much better if the woman takes care of the home. All differences are statistically significant.

As expected, the more educated people are, the more likely they are to approve of nontraditional gender roles. Those with less than a high school degree are the most likely to believe women should primarily take care of the home (56.3 percent) while those with a high school education (42.6 percent) and those who have gone to college (29.6 percent) are the least likely to hold this attitude. The results are statistically significant. The hypothesis is supported.

How about race? Are there racial differences in this measure of gender role attitudes?

Data File: **GSS**
Task: **Auto-Analyzer**
Variable: **66) WIFE@HOME**
➤ View: **Race**

WIFE@HOME -- AGREE OR DISAGREE?: It is much better for everyone involved if the man is the achiever outside the home and the woman takes care of the home and family. (FEFAM)

	WHITE	BLACK
AGREE	38.6%	34.8%
DISAGREE	61.4%	65.2%
Number of cases	2854	549

African-Americans and whites are about equally likely to think it is much better if the woman takes care of the home.

According to the summary below this table, African Americans (34.8 percent) and whites (38.6 percent) are about equally as likely to believe that it is best for women to take care of the home. No, there is not a statistically significant difference between how blacks and whites answer this question.

Another comparison we can make is between those with different religious preferences.

Data File: **GSS**
Task: **Auto-Analyzer**
Variable: **66) WIFE@HOME**
➤ View: **Religion**

WIFE@HOME -- AGREE OR DISAGREE?: It is much better for everyone involved if the man is the achiever outside the home and the woman takes care of the home and family. (FEFAM)

	LIB. PROT.	CON.PROT.	CATHOLIC	JEWISH	NONE
AGREE	35.5%	49.5%	38.7%	26.1%	19.3%
DISAGREE	64.5%	50.5%	61.3%	73.9%	80.7%
Number of cases	747	1042	875	69	513

Conservative Protestants are most likely (49.5%) to think it is much better if the woman takes care of the home and those reporting no religion (19.3%) are least likely. The difference is statistically significant.

What we see here is that most people, regardless of their religious orientation, approve of women working outside the home. However, there are some statistically significant differences. Conservative Protestants are the most likely (49.5 percent) to believe that women should put home and family first, while those who are not religious are the least likely (19.3 percent) to agree. Liberal Protestants (35.5 percent) and Catholics (38.7 percent) fall midway in between.

Finally, in many recent elections, a lot of attention has been given to the *gender gap*—the tendency for women to vote for Democratic rather than Republican presidential candidates. So, are Democrats more likely to approve of women working outside the home?

Data File: **GSS**
Task: **Auto-Analyzer**
Variable: **66) WIFE@HOME**
➤ View: **Party**

WIFE@HOME -- AGREE OR DISAGREE?: It is much better for everyone involved if the man is the achiever outside the home and the woman takes care of the home and family. (FEFAM)

	DEMOCRAT	INDEPEND.	REPUBLICAN
AGREE	34.4%	37.0%	43.8%
DISAGREE	65.6%	63.0%	56.2%
Number of cases	1669	692	1196

Republicans are most likely (43.8%) to think it is much better if the woman takes care of the home and Democrats (34.4%) are least likely. The difference is statistically significant.

Based on these results, Republicans are the most likely to agree that it is best for all concerned if women take care of the home and family (43.8 percent). Democrats (34.4 percent) and Independents (37 percent) are less likely to hold this attitude.

The variation in gender role attitudes we have seen across these different demographic categories illustrates how the social environment can influence the beliefs we hold. Our age, religion, education, and even our political party affiliation have the potential to influence our gender role attitudes. But gender roles are not just passively shaped by society; our attitudes also have the potential to influence our social environment and to guide other decisions we make. Gender role attitudes influence whether or not to marry and who to marry, whether or not to go to college, which church to attend, which career to pursue, and many other important decisions that we all make.

Gender role attitudes don't exist in a vacuum; they are a part of the world in which we live. For example, a person who is doing the hiring for a company may discriminate against women or men because of his or her own gender role attitudes. Let's take a look at the extent to which men and women believe that women are likely to face discrimination in the workplace.

Data File: **GSS**
➤ Task: **Cross-tabulation**
➤ Row Variable: **71) DISC WOMAN**
➤ Column Variable: **25) SEX**
➤ View: **Tables**
➤ Display: **Column %**

DISC WOMAN by SEX
Cramer's V: 0.126 **

DISC WOMAN	SEX: MALE	FEMALE	TOTAL
LIKELY	218	339	557
	65.9%	77.2%	72.3%
UNLIKELY	113	100	213
	34.1%	22.8%	27.7%
Missing	2130	2749	4879
TOTAL	331	439	770
	100.0%	100.0%	

A majority of both men and women believe it is likely that a woman won't get a job or a promotion even if she is equally, or even more, qualified for the position than a man. There is a statistically significant difference between women (77.2 percent) and men (65.9 percent) on this issue, but both sexes clearly believe that such discrimination is likely to occur.

Of course, some would argue that men can also be discriminated against in the workplace. Thus, the GSS asked "What do you think the chances are these days that a man won't get a job or promotion while an equally or less qualified woman gets one instead?"

		Data File:	**GSS**
		Task:	**Cross-tabulation**
➤	Row Variable:	**70) DISC MAN**
➤	Column Variable:	**25) SEX**
➤	View:	**Tables**
➤	Display:	**Column %**

DISC MAN   by   SEX
Cramer's V: 0.100 **

		SEX		
		MALE	FEMALE	TOTAL
DISC MAN	LIKELY	201	218	419
		55.4%	45.2%	49.6%
	UNLIKELY	162	264	426
		44.6%	54.8%	50.4%
	Missing	2098	2706	4804
	TOTAL	363	482	845
		100.0%	100.0%	

The respondents are divided more evenly with regard to the issue of discrimination against men. Again, there is a small, but statistically significant, difference between men and women. But this time it is the men (55.4 percent) who are more likely than the women (45.2 percent) to say that this type of discrimination is likely to occur (V = .100**).

Perceptions of discrimination do appear to vary depending upon whether or not the person being asked is the target of the type of discrimination in question. Women are slightly more likely to believe that women are discriminated against, and men are more likely to believe that men are discriminated against. Overall, however, both men and women agree that sex discrimination is more common against women than it is against men.

One of the areas where we would expect to see evidence of any sex discrimination would be in a person's income. Let's do an income comparison of male respondents with female respondents using the subset of those who are employed full-time.

		Data File:	**GSS**
		Task:	**Cross-tabulation**
➤	Row Variable:	**29) R.INCOME**
➤	Column Variable:	**25) SEX**
➤	Subset Variable:	**30) WORK STAT**
➤	Subset Category:	**Include: 1) FULL TIME**
➤	View:	**Tables**
➤	Display:	**Column %**

R.INCOME   by   SEX

Cramer's V: 0.249 **

		SEX		
		MALE	FEMALE	TOTAL
R.INCOME	$0K-17.4K	209	380	589
		14.5%	29.3%	21.5%
	17.5K-34.9	520	572	1092
		36.2%	44.1%	39.9%
	$35K +	708	346	1054
		49.3%	26.7%	38.5%
	Missing	195	162	357
	TOTAL	1437	1298	2735
		100.0%	100.0%	

The option for selecting a subset variable is located on the same screen you use to select other variables. For this example, select 30) WORK STAT as a subset variable. A window will appear that shows you the categories of the subset variable. Select 1) FULL TIME as your subset category and choose the [Include] option. Then click [OK] and continue as usual.

With this particular subset selected, the results will be limited to those in the sample who are currently employed full-time. The subset selection continues until you exit the task or clear the subset variable.

There, in fact, is a large income discrepancy between males and females in the General Social Survey. Proportionally, there are twice as many women (29.3 percent) than men (14.5 percent) in the lowest income category. And there are nearly twice as many men (49.3 percent) than women (26.7 percent) in the highest income category. Women who are employed full-time tend to be paid much less than men who are employed full-time (V = 0.249**).

Let's shift from looking at gender role attitudes to looking at actual role enactments.

Data File: **GSS**
➤ Task: **Univariate**
➤ Primary Variable: **32) MAR.F.WRK**
➤ View: **Pie**

MAR.F.WRK -- Work status of married females

		Freq.	%
■	0) FULL TIME	646	52.5
▨	1) PART-TIME	211	17.1
▩	2) KEEP HOUSE	374	30.4
	TOTAL (N)	1231	100.0
	Missing	4418	

This variable, MAR.F.WRK (which is short for *married women's work status*), divides married women into three categories based on their level of employment. (Married women who are retired or are full-time students are not included in this sample.) In this sample, 52.5 percent of the married women are employed full-time outside the home, and another 17.1 percent are employed part-time. Just under one-third (30.4 percent) of the married women surveyed are not employed outside the home. In the worksheet section that follows, you will have the opportunity to examine some of the differences between the women in each of these three groups.

WORKSHEET

NAME: _____

COURSE: _____

DATE: _____

CHAPTER
3

REVIEW QUESTIONS

Based on the first part of this exercise, answer True or False to the following items:

In general, the level of emphasis a society places on boys being tough tends to be similar to the emphasis the society places on girls being tough.	T F
In about half of preindustrial societies, housework was never done by men.	T F
For the most part, attitudes toward household chores and the role of women are similar across modern-day nations.	T F
A majority of both men and women approve of women being employed outside the home.	T F
No matter how you ask the question, men and women have the same attitudes toward women working outside the home.	T F
The more education one has, the less likely one is to believe that women should take care of the home and family.	T F
A majority of men believe that women are likely to be discriminated against in the workplace.	T F
Less than one-third of the married women in the GSS are employed full-time.	T F

EXPLORIT QUESTIONS

1. The hypothesis is: **Women who attend conservative churches will be less likely than those who attend moderate or liberal churches to be employed full-time.**

> ➤ *Data File:* **GSS**
> ➤ *Task:* **Cross-tabulation**
> ➤ *Row Variable:* **32) MAR.F.WRK**
> ➤ *Column Variable:* **48) RELIGION**
> ➤ *View:* **Tables**
> ➤ *Display:* **Column %**

   a. Fill in the percentaged results for the *top* row of this table.

Chapter 3: Gender Roles

	LIB. PROT.	CON. PROT.	CATHOLIC	JEWISH
FULL TIME	_____%	_____%	_____%	_____%

b. What is the value of V? $V =$ _____

c. Is V statistically significant? Yes   No

d. Is the hypothesis supported? Yes   No

e. In the first part of this chapter, we saw that conservative Protestants were less likely to approve of women working outside the home. Therefore, what would you conclude based on the results of this hypothesis? (circle one)

    1. Most women who are conservative Protestants choose not to be employed.

    2. Many women who are conservative Protestants may find it necessary to work outside the home although they would prefer not to be employed.

    3. Most women who are conservative Protestants want to be employed but cannot find a job.

2. The hypothesis is: **Women with more education will be more likely to be employed full-time.**

             Data File: **GSS**
                 Task: **Cross-tabulation**
       Row Variable: **32) MAR.F.WRK**
   ➤ *Column Variable:*   **27) EDUCATION**
           ➤ *View:*   **Tables**
        ➤ *Display:*   **Column %**

a. Fill in the percentaged results for the *top* row of this table.

	NO HS GRAD	HS GRAD	COLL EDUC
FULL TIME	_____%	_____%	_____%

b. What is the value of V? $V =$ _____

c. Is V statistically significant? Yes   No

d. Is the hypothesis supported? Yes   No

e. Which group is the least likely to be employed full-time? (circle one)

    1. married women who did not graduate from high school

    2. married women with a high school degree who did not go to college

    3. married women who went to college

f. Based on these results, one would most likely conclude that those who are least likely to be employed (circle one)
1. are better off financially and therefore do not need to work outside the home.
2. do not have the necessary skills or the credentials to find a good full-time job.
3. have the skills and credentials to find a good full-time job but choose not to be employed.

3. Does whether or not a woman has children living at home influence her work status?

> Data File: **GSS**
> Task: **Cross-tabulation**
> Row Variable: **32) MAR.F.WRK**
> ➤ Column Variable: **16) KIDS<18**
> ➤ View: **Tables**
> ➤ Display: **Column %**

a. Fill in the percentaged results for this table.

	NO	YES
FULL TIME	_____ %	_____ %
PART-TIME	_____ %	_____ %
KEEP HOUSE	_____ %	_____ %

b. What is the value of V?

V = _____

c. Is V statistically significant?

Yes   No

d. Most married women with children keep house full-time.

T   F

e. Married women with children are less likely to keep house full-time.

T   F

4. If the children living at home are under six years of age, how does that affect the work status of a married woman?

> Data File: **GSS**
> Task: **Cross-tabulation**
> Row Variable: **32) MAR.F.WRK**
> ➤ Column Variable: **17) KIDS<6**
> ➤ View: **Tables**
> ➤ Display: **Column %**

Chapter 3: Gender Roles

a. Fill in the percentaged results for this table.

	NO	YES
FULL TIME	_____%	_____%
PART-TIME	_____%	_____%
KEEP HOUSE	_____%	_____%

b. What is the value of V?   V = _____

c. Is V statistically significant?   Yes   No

d. Most married women with young children keep house full-time.   T   F

e. Married women with young children are less likely to keep house full-time.   T   F

5. Next, examine the relationship between a woman's work status and marital happiness.

> Data File: **GSS**
> Task: **Cross-tabulation**
> ➤ Row Variable: **74) HAP.MARR.?**
> ➤ Column Variable: **32) MAR.F.WRK**
> ➤ View: **Tables**
> ➤ Display: **Column %**

Which of the following statements best summarizes these results? (circle one)

     a. Women who are employed full-time tend to have the happiest marriages.

     b. Women who are employed part-time tend to have the happiest marriages.

     c. Women who are full-time homemakers tend to have the happiest marriages.

     d. There is not a statistically significant relationship between a woman's work status and marital happiness.

6. What percentage of survey respondents' mothers worked for at least one year while they were growing up? Select the appropriate statistical task and use 46) MA WRK GRW as your variable.

a. Yes, mother worked   _____%

b. No, mother didn't work   _____%

7. Do children whose mothers were employed believe that it affected the mother-child relationship?

>   Data File: **GSS**
>   Task: **Cross-tabulation**
>   ➤ Row Variable: **64) MOTH.WORK?**
> ➤ Column Variable: **46) MA WRK GRW**
>   ➤ View: **Tables**
>   ➤ Display: **Column %**

In a brief paragraph, summarize the relationship in this table. Remember to indicate the strength of the relationship and to state whether the results are statistically significant. (It's important that you read the complete descriptions for both variables before you start writing.)

8. Let's return to the GLOBAL file and examine how females differ from males in terms of education and labor-force participation. The variable 33) M/F EDUC. is the average years of female schooling as a percentage of the average years of male schooling. In short, a higher value for a nation indicates that female education is more similar to or greater than male education.

The hypothesis is: **Nations with proportionally higher female education rates will tend to have higher percentages of women employed in professional or technical jobs.**

>   ➤ Data File: **GLOBAL**
>   ➤ Task: **Scatterplot**
>   ➤ Dependent Variable: **30) FEM.PROF.**
> ➤ Independent Variable: **33) M/F EDUC.**
>   ➤ View: **Reg. Line**

a. What is the correlation coefficient?   r = _____

b. Is it statistically significant?   Yes   No

c. Do these results support the hypothesis?   Yes   No

9. Let's test two hypotheses. **Hypothesis 1: Nations that have high proportions of employed women will have higher percentages of women who hold political office. Hypothesis 2: Nations with high rates of women working in professional or technical jobs will have higher percentages of women who hold political office.** (Note: Carefully read the complete descriptions for these variables before you do your analysis.)

> *Data File:* **GLOBAL**
> ➤ *Task:* **Correlation**
> ➤ *Variables:* **30) FEM.PROF.**
>   **73) F/M EMPLOY**
>   **74) %FEM.HEADS**
>   **31) %FEM.LEGIS**

    a. As the female employment rate increases, so does the percentage of parliamentary seats held by females.   T   F

    b. As the female employment rate increases, so does the percentage of top-level political seats (e.g., ministerial level, presidential cabinet) held by females.   T   F

    c. As the rate of women working in professional or technical jobs increases, so does the percentage of top-level political seats (e.g., ministerial level, presidential cabinet) held by females.   T   F

    d. If you were trying to guess whether a particular nation has a high percentage of women who hold political office, would it be better to know the nation's female employment rate or the percentage of women who hold professional jobs? (circle one)

       1. Female employment rate

       2. Percentage of women with professional jobs

CHAPTER 4

PREMARITAL SEX: ATTITUDES AND BEHAVIOR

Tasks: Univariate, Historical Trends, Cross-tabulation, Auto-Analyzer, Mapping
Data Files: GSS, CULTURES, TRENDS, GLOBAL

Sociologist Ira Reiss has described four moral standards of premarital sexuality. The first is the *abstinence standard*, which is the belief that premarital sexual intercourse is wrong for both men and women, regardless of their feelings for each other. This was the most commonly held sexual standard in the United States until the 1950s or 1960s. The second is what Reiss calls the *double standard*, or the belief that premarital intercourse is permissible for men but not for women. Although not widely discussed publicly, this has been a very common standard until perhaps just recently. The third is *permissiveness with affection*, which is the belief that premarital intercourse is acceptable as long as the couple has an affectionate, stable relationship. *Permissiveness with or without affection* is the fourth standard. As the name suggests, this is the belief that premarital sex is never wrong as long as it is consensual. Which of these beliefs do you hold? We do not have a U.S. data set that will allow us to clearly examine each of these four standards, but we can use some other questions to arrive at an understanding of what Americans believe about premarital sexual intercourse. Along the way, we will look for variations in the beliefs held by different groups.

> ➤ *Data File:* **GSS**
> ➤ *Task:* **Univariate**
> ➤ *Primary Variable:* **88) PREM.SEX**
> ➤ *View:* **Pie**

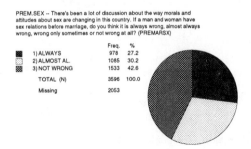

PREM.SEX -- There's been a lot of discussion about the way morals and attitudes about sex are changing in this country. If a man and woman have sex relations before marriage, do you think it is always wrong, almost always wrong, wrong only sometimes or not wrong at all? (PREMARSX)

		Freq.	%
■	1) ALWAYS	978	27.2
▨	2) ALMOST AL.	1085	30.2
▦	3) NOT WRONG	1533	42.6
	TOTAL (N)	3596	100.0
	Missing	2053	

The responses are fairly divided, but the largest group consists of those who believe premarital sex is not wrong (42.6 percent). This group would probably express a view similar to Reiss' *permissiveness with or without affection* moral standard. Just over one-fourth of the respondents believe premarital sex is always wrong (27.2 percent). These respondents would be in the abstinence category. The rest of the respondents, 30.2 percent, believe premarital sex is almost always wrong. These respondents are probably closest to Reiss' *permissiveness with affection* standard.

How do people feel about young teens having premarital sex?

Data File: **GSS**
Task: **Univariate**
➤ Primary Variable: **89) TEEN SEX**
➤ View: **Pie**

TEEN SEX -- What if they are in their early teens, say 14 to 16 years old? In that case, do you think sex relations before marriage are always wrong, almost always wrong, wrong only sometimes, or not wrong at all? (TEENSEX)

		Freq.	%
■	1) ALWAYS	2655	72.1
▨	2) ALMOST AL.	900	24.4
▦	3) NOT WRONG	128	3.5
	TOTAL (N)	3683	100.0
	Missing	1966	

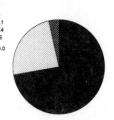

By comparison with the prior results that showed 27.2 percent of adult Americans believe that premarital sex for people of all ages is always wrong, 72.1 percent of those surveyed believe that premarital sex is always wrong for those in their early teens. Only 3.5 percent of the respondents believe that premarital sex for this age group is never wrong, by comparison with the 42.6 percent who would say that about premarital sex in general. Thus, premarital sexual standards in general are fairly permissive, but most people believe it is better for those in their young teens to not engage in sexual activity.

Before we continue our examination of U.S. attitudes toward premarital sex, it can be helpful to step back and look at this topic from a cross-cultural perspective. You'll recall from Chapter 1 that the CULTURES file is based on data that was coded from detailed ethnographies and reports written decades or even centuries ago. Few of the preindustrial societies on which these data are based still exist today, and some have not existed for centuries. Moreover, we don't have survey data for these societies with which to examine attitudes toward premarital sex. But for many of these societies, we do have data indicating practice and acceptance of premarital sexual activity.

➤ Data File: **CULTURES**
➤ Task: **Univariate**
➤ Primary Variable: **67) TEEN SEX?**
➤ View: **Pie**

TEEN SEX? -- IS PREMARITAL SEX CONDONED? (MGH)

		Freq.	%
■	0) NO	29	44.6
▨	1) YES	36	55.4
	TOTAL (N)	65	100.0
	Missing	121	

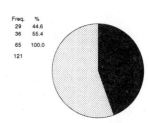

Over half (55.4 percent) of preindustrial societies condoned premarital sex. Let's follow up on Ira Reiss' second standard of premarital sex: the *double standard*. Is this a phenomenon unique to the United States, or did this occur in preindustrial societies across the world?

Data File: **CULTURES**

Task: **Univariate**

➤ Primary Variable: **63) PREMARITAL**

➤ View: **Pie**

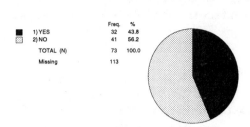

A double standard is evident in almost 44 percent of the societies included here. But it's not entirely clear from this coding what the double standard is. Is premarital sex more permissible for men than for women? Or are there societies in which premarital sex is more permissible for women? Also, it seems that if premarital sex is very permissible for one sex, it must be at least somewhat permissible for the other. Let's begin by examining the sexual behavior of boys, and then girls.

Data File: **CULTURES**

Task: **Univariate**

➤ Primary Variable: **58) SEX (BOYS)**

➤ View: **Pie**

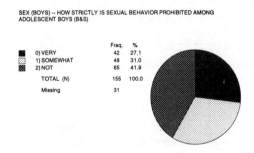

Sexual behavior among adolescent boys is strictly prohibited in 27.1 percent of societies and is somewhat prohibited in 31.0 percent of societies.

Data File: **CULTURES**

Task: **Univariate**

➤ Primary Variable: **60) SEX(GIRLS)**

➤ View: **Pie**

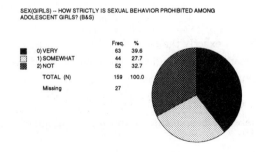

And for girls, sexual behavior is strictly prohibited in 39.6 percent of societies and is somewhat prohibited in 27.7 percent of societies. Or, to look at it another way, 41.9 percent of societies do not restrict sexual behavior for adolescent boys, compared to 32.7 percent of societies for adolescent girls. Based on these results, it appears that a double standard may exist.

This data file includes versions of these same two variables in which the somewhat strict and not strict categories have been combined. Let's cross-tabulate these variables to see whether societies that permit sexual behavior for boys are the same societies that permit sexual behavior for girls.

Data File: **CULTURES**
➤ Task: **Cross-tabulation**
➤ Row Variable: **61) SEX(GIRL)***
➤ Column Variable: **59) SEX(BOYS)***
➤ View: **Tables**
➤ Display: **Column %**

SEX(GIRL)*   by   SEX(BOYS)*
Cramer's V: 0.777 **

		SEX(BOYS)*			
		NOT STRICT	STRICTLY	Missing	TOTAL
SEX(GIRL)*	NOT STRICT	93	1	2	94
		86.1%	2.4%		62.7%
	STRICTLY	15	41	7	56
		13.9%	97.6%		37.3%
	Missing	5	0	22	27
	TOTAL	108	42	31	150
		100.0%	100.0%		

These results are interesting, although not entirely surprising. In societies that strictly prohibit sexual behavior of boys, all but one society also strictly prohibits sexual behavior by adolescent girls. However, in societies where sexual behavior by adolescent boys is not strictly prohibited, 13.9 percent of these same societies have strict prohibitions for girls. Clearly, a double standard does exist in some societies.

Let's examine this from another angle. We would expect to find a double standard in premarital sex in societies where sexual behavior of girls is strictly forbidden.

Data File: **CULTURES**
Task: **Cross-tabulation**
➤ Row Variable: **63) PREMARITAL**
➤ Column Variable: **61) SEX(GIRL)***
➤ View: **Tables**
➤ Display: **Column %**

PREMARITAL   by   SEX(GIRL)*
Cramer's V: 0.367 **

		SEX(GIRL)*			
		NOT STRICT	STRICTLY	Missing	TOTAL
PREMARITAL	YES	13	18	1	31
		31.7%	69.2%		46.3%
	NO	28	8	5	36
		68.3%	30.8%		53.7%
	Missing	55	37	21	113
	TOTAL	41	26	27	67
		100.0%	100.0%		

This is exactly the case. Societies that strictly prohibit sexual behavior in girls are more likely (69.2 percent) to have a double standard toward premarital sex than societies that don't strictly prohibit sexual behavior in girls (31.7 percent).

Now, let's return to the United States and look at attitudes over the past couple of decades to see if there has been a change in the percentage of Americans who believe premarital sex is always wrong.

➤ *Data File:* **TRENDS**
  ➤ *Task:* **Historical Trends**
➤ *Variable:* **22) PREM.SEX%**

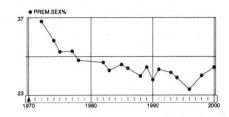

Percentage who believe premarital sex is always wrong

Looking at this line graph we see that the percentage of respondents who believe premarital sex is always wrong declines for most years from 1972 through the mid-1980s, then levels off before dropping sharply again in 1996. Although there is a slight rebound from 1996 to 2000, overall there has been movement away from the abstinence standard over the last few decades.

Another way we can look for trends in sexual values is by examining the relationship between age and attitudes toward premarital sex. If our society is moving toward more permissive sexual standards, then **those who are older will be more likely than those who are younger to believe that premarital sex is always wrong**.

➤ *Data File:* **GSS**
  ➤ *Task:* **Cross-tabulation**
➤ *Row Variable:* **88) PREM.SEX**
➤ *Column Variable:* **24) AGE**
  ➤ *View:* **Tables**
  ➤ *Display:* **Column %**

PREM.SEX   by   AGE
Cramer's V: 0.153 **

		AGE				
		<30	30–49	50 AND UP	Missing	TOTAL
P R E M . S E X	ALWAYS	121	384	473	0	978
		18.2%	23.3%	37.2%		27.3%
	ALMOST AL.	197	455	427	6	1079
		29.6%	27.6%	33.6%		30.1%
	NOT WRONG	347	811	372	3	1530
		52.2%	49.2%	29.2%		42.7%
	Missing	398	861	791	3	2053
	TOTAL	665	1650	1272	12	3587
		100.0%	100.0%	100.0%		

As expected, those under age 30 (18.2 percent) are the least likely to believe that premarital sex is always wrong and those over age 50 (37.2 percent) are the most likely to hold this belief. The differences between most of the age categories are statistically significant. Either our society is moving toward more permissive sexual standards, or people tend to adopt more conservative standards as they grow older. Given the historical trend we looked at earlier, the first explanation would seem to be the more likely.

Let's try to explain more of the variation in attitudes toward premarital sex, beginning with the hypothesis that **people who have never been married will be less likely to believe that premarital sex is always wrong**. The independent variable in this hypothesis is marital status; the dependent variable is attitude toward premarital sex.

Data File: **GSS**
Task: **Cross-tabulation**
Row Variable: **88) PREM.SEX**
➤ Column Variable: **3) EVER MAR?**
➤ View: **Tables**
➤ Display: **Column %**

PREM.SEX   by   EVER MAR?
Cramer's V: 0.164 **

		EVER MAR?			
		YES	NO	Missing	TOTAL
P R E M . S E X	ALWAYS	834	143	1	977
		30.7%	16.3%		27.2%
	ALMOST AL	838	247	0	1085
		30.8%	28.2%		30.2%
	NOT WRONG	1046	487	0	1533
		38.5%	55.5%		42.6%
	Missing	1554	498	1	2053
	TOTAL	2718	877	2	3595
		100.0%	100.0%		

As we read across the top row in this table, we see that 30.7 percent of those who have been married hold this belief, while 16.3 percent of those who have never married believe premarital sex is always wrong. Since this relationship is moderately strong, as indicated by Cramer's V (V = .164**), and statistically significant, we can say the results are consistent with the stated hypothesis.

It is commonly believed that men are more willing than women to engage in premarital sex. Does this mean that fewer men believe premarital sex is wrong?

The hypothesis is: **Men are less likely than women to believe that premarital sex is always wrong.**

Data File: **GSS**
Task: **Cross-tabulation**
Row Variable: **88) PREM.SEX**
➤ Column Variable: **25) SEX**
➤ View: **Tables**
➤ Display: **Column %**

PREM.SEX   by   SEX
Cramer's V: 0.103 **

		SEX		
		MALE	FEMALE	TOTAL
P R E M . S E X	ALWAYS	367	611	978
		23.5%	30.0%	27.2%
	ALMOST AL	441	644	1085
		28.2%	31.7%	30.2%
	NOT WRONG	754	779	1533
		48.3%	38.3%	42.6%
	Missing	899	1154	2053
	TOTAL	1562	2034	3596
		100.0%	100.0%	

The direction of the findings is consistent with the hypothesis: 23.5 percent of the males believe premarital sex is always wrong compared to 30 percent of the females. The difference is statistically significant and the hypothesis is supported. Men are less disapproving of premarital sex than are women.

The three institutions that play the most direct role in shaping our personal values are the family, the church, and the school. The extent to which any of these institutions deliberately teach sexual values varies greatly. There also is a significant amount of variation within institutions. For example, not all parents will discuss sexual issues with their children, and the values that are taught will not all be the same. The same is true for churches and schools: both the amount of teaching that takes place and the material that is taught will vary from one church or school to the next. Can some of the variation in attitudes toward premarital sex be explained by taking into account the influence of these three social institutions?

Earlier it was demonstrated that a respondent's current marital status influences his or her attitude toward premarital sex—but what about the structure of the family that person was socialized into? Is being raised in a two-parent home, as opposed to a one-parent home, a predictor of sexual attitudes? The variable FAMILY @16 is based on a question that asked respondents whether or not they lived with both parents at the age of 16.

Data File: **GSS**
Task: **Cross-tabulation**
Row Variable: **88) PREM.SEX**
➤ Column Variable: **45) FAMILY @16**
➤ View: **Tables**
➤ Display: **Column %**

PREM.SEX   by   FAMILY @16
Cramer's V: 0.058 **

		FAMILY @16		
		YES	NO	TOTAL
P R E M . S E X	ALWAYS	707	271	978
		28.6%	24.0%	27.2%
	ALMOST AL.	753	332	1085
		30.5%	29.5%	30.2%
	NOT WRONG	1009	524	1533
		40.9%	46.5%	42.6%
	Missing	1426	627	2053
	TOTAL	2469	1127	3596
		100.0%	100.0%	

There doesn't seem to be much difference between these two groups. Reading across the bottom row of the table, we see that 40.9 percent of those who lived with both parents at age 16 think premarital sex is not wrong. This compares to 46.5 percent of those who did not live with both parents at that age. Although the difference is statistically significant, the value of V (V = .058**) denotes that this is a very weak relationship. Being raised in a two-parent home is not a good predictor of sexual attitudes.

The second agent of socialization to explore is the church. The hypothesis is: **People who attend church on a regular basis will be more likely to believe that premarital sex is always wrong**.

Data File: **GSS**
Task: **Cross-tabulation**
Row Variable: **88) PREM.SEX**
➤ Column Variable: **50) ATTEND**
➤ View: **Tables**
➤ Display: **Column %**

PREM.SEX   by   ATTEND
Cramer's V: 0.318 **

		ATTEND				
		NEVER	MONTH/YRLY	WEEKLY	Missing	TOTAL
P R E M . S E X	ALWAYS	93	282	585	18	960
		12.7%	16.2%	55.3%		27.2%
	ALMOST AL.	176	600	286	23	1062
		24.0%	34.5%	27.1%		30.1%
	NOT WRONG	465	857	186	25	1508
		63.4%	49.3%	17.6%		42.7%
	Missing	395	948	652	58	2053
	TOTAL	734	1739	1057	124	3530
		100.0%	100.0%	100.0%		

The largest difference in this table is between those who attend church weekly and the rest of the categories. Over half, 55.3 percent, of the weekly attendees believe premarital sex is always wrong. This is more than three times the percentage for those who attend monthly or annually (16.2 percent), and more than four times the percentage for those who never attend church (12.7 percent). The relationship is very strong (V = .318**). Hence, the hypothesis is supported.

Let's explore the influence of religion a little further to see if the type of church people attend affects what they believe about premarital sex.

Data File:	**GSS**
Task:	**Cross-tabulation**
Row Variable:	**88) PREM.SEX**
➤ Column Variable:	**48) RELIGION**
➤ View:	**Tables**
➤ Display:	**Column %**

PREM.SEX   by   RELIGION
Cramer's V: 0.266 **

		RELIGION						
		LIB. PROT.	CON.PROT.	CATHOLIC	JEWISH	NONE	Missing	TOTAL
P R E M . S E X	ALWAYS	171	486	157	4	42	118	860
		23.3%	46.9%	18.3%	5.8%	8.0%		26.7%
	ALMOST AL	270	272	296	25	117	105	980
		36.7%	26.2%	34.6%	36.2%	22.3%		30.4%
	NOT WRONG	294	279	403	40	366	151	1382
		40.0%	26.9%	47.1%	58.0%	69.7%		42.9%
	Missing	413	587	528	44	269	212	2053
	TOTAL	735	1037	856	69	525	586	3222
		100.0%	100.0%	100.0%	100.0%	100.0%		

Religious affiliation does make a difference. Conservative Protestants (46.9 percent) are more than twice as likely as either liberal Protestants (23.3 percent) or Catholics (18.3 percent) to believe that premarital sex is always wrong. Only 5.8 percent of those who are Jewish and 8 percent of those with no religious affiliation hold this belief. These results are statistically significant (V = .266**). So, both church attendance and the type of church one attends are significant predictors of attitudes about premarital sex.

The third institution that often contributes directly to sexual socialization is the school. Because educational institutions tend to encourage students to challenge traditional ideas and norms, we will hypothesize that **the more educated a person is, the less likely he or she is to believe that premarital sex is always wrong**.

Data File:	**GSS**
Task:	**Cross-tabulation**
Row Variable:	**88) PREM.SEX**
➤ Column Variable:	**27) EDUCATION**
➤ View:	**Tables**
➤ Display:	**Column %**

PREM.SEX   by   EDUCATION
Cramer's V: 0.071 **

		EDUCATION				
		NO HS GRAD	HS GRAD	COLL.EDUC.	Missing	TOTAL
P R E M . S E X	ALWAYS	209	312	451	6	972
		34.3%	30.1%	23.3%		27.1%
	ALMOST AL.	175	311	596	3	1082
		28.7%	30.0%	30.8%		30.2%
	NOT WRONG	226	415	886	6	1527
		37.0%	40.0%	45.8%		42.6%
	Missing	366	636	1045	6	2053
	TOTAL	610	1038	1933	21	3581
		100.0%	100.0%	100.0%		

Overall, as the level of education rises, the percentage who disapprove of premarital sex declines. The biggest difference is between those who did not go to college and those who did. Among those who went to college, only 23.3 percent believe premarital sex is always wrong, compared to 34.3 percent for those who did not graduate high school and 30.1 percent for those with a high school diploma. The hypothesis is supported—higher education is associated with more permissive attitudes about premarital sex.

Of course, not everybody believes that sex education should be a part of the school curriculum. Some believe that sex education is something that should be taught by the parents. As public institutions, most schools have traditionally avoided making value judgments on sexual behavior. Thus, it seems likely that **those who believe that premarital sex is always wrong will be the least likely to support sex education in schools**.

Data File: **GSS**
Task: **Cross-tabulation**
➤ Row Variable: **87) SEX ED?**
➤ Column Variable: **88) PREM. SEX**
➤ View: **Tables**
➤ Display: **Column %**

SEX ED?   by   PREM.SEX
Cramer's V: 0.307 **

		PREM.SEX				
		ALWAYS	ALMOST AL.	NOT WRONG	Missing	TOTAL
SEX ED?	FOR	659	947	1450	121	3056
		70.7%	89.3%	95.8%		87.2%
	AGAINST	273	114	63	18	450
		29.3%	10.7%	4.2%		12.8%
	Missing	46	24	20	1914	2004
	TOTAL	932	1061	1513	2053	3506
		100.0%	100.0%	100.0%		

Note that you must make 88) PREM.SEX the column variable because it is the independent variable in this hypothesis.

A majority of the respondents in each of the categories favor sex education in public schools, but there are large differences between the categories based upon the respondents' own sexual values. Among those who believe premarital sex is always wrong, 70.7 percent are for sex education in public schools, as compared to 95.8 percent for those who believe premarital sex is not wrong. This relationship is strong and statistically significant (V = .307**), so the hypothesis is supported.

In examining the variations in attitudes toward premarital sex, we have found that current marital status, gender, church attendance, and education are all significant predictors of premarital sexual attitudes in varying degrees. The dependent variable in each of the preceding tables was one's attitude toward premarital sex in general. Now, let's explore whether some of these same factors influence attitudes toward homosexuality. This time we will use ExplorIt's AUTO-ANALYZER task.

Data File: **GSS**
➤ Task: **Auto-Analyzer**
➤ Variable: **91) HOMO.SEX**
➤ View: **Univariate**

HOMO.SEX -- What about sexual relations between two adults of the same sex -- do you think it is always wrong, almost always wrong, wrong only sometimes, or not wrong at all? (HOMOSEX)

	%
ALWAYS	58.4%
ALMOST AL.	12.5%
NOT WRONG	29.1%
Number of cases	3450

Among all respondents, 58.4% of the sample think sex between homosexual people is always wrong.

This first view gives us a look at the distribution of responses for the HOMO.SEX variable. Overall, 58.4 percent of those surveyed believe homosexuality is always wrong, 12.5 percent believe it is almost always wrong, and the remaining 29.1 percent believe homosexuality is not at all wrong. Now let's break these percentages down using some of the demographic variables—beginning with age.

Data File: **GSS**
Task: **Auto-Analyzer**
Variable: **91) HOMO.SEX**
➤ View: **Age**

HOMO.SEX -- What about sexual relations between two adults of the same sex -- do you think it is always wrong, almost always wrong, wrong only sometimes, or not wrong at all? (HOMOSEX)

	<30	30-49	50 AND UP
ALWAYS	44.4%	55.2%	69.3%
ALMOST AL.	12.2%	14.0%	10.9%
NOT WRONG	43.5%	30.8%	19.8%
Number of cases	658	1507	1279

The higher the age, the more likely the individual is to think sex between homosexual people is always wrong. All differences are statistically significant.

The higher the age, the less likely the individual is to think sex between homosexual people is not wrong. All differences are statistically significant.

The percentage of respondents who believe that homosexuality is always wrong increases as the age category increases. Those under 30 (44.4 percent) are the least likely, and those over 50 are the most likely (69.3 percent), to believe homosexuality is always wrong. Respondents age 30–49 fall midway in between these two values (55.2 percent). Conversely, the percentage of those who believe homosexuality is not at all wrong decreases from 43.5 percent for those in the youngest age category to 19.8 for those in the oldest age category. The summary below the table tells us that these differences are statistically significant. These results suggest that there is an increasing acceptance of homosexuality in the United States.

Data File: **GSS**
Task: **Auto-Analyzer**
Variable: **91) HOMO.SEX**
➤ View: **Education**

HOMO.SEX -- What about sexual relations between two adults of the same sex -- do you think it is always wrong, almost always wrong, wrong only sometimes, or not wrong at all? (HOMOSEX)

	NO HS GRAD	HS GRAD	COLL EDUC
ALWAYS	74.5%	65.4%	48.9%
ALMOST AL.	7.2%	11.5%	14.9%
NOT WRONG	18.2%	23.1%	36.2%
Number of cases	609	1032	1800

The higher the education, the less likely the individual is to think sex between homosexual people is always wrong. All differences are statistically significant.

The higher the education, the more likely the individual is to think sex between homosexual people is not wrong. All differences are statistically significant.

Here we see that the higher the education, the less likely the respondent is to think that homosexuality is always wrong. These results are similar to the earlier analysis of attitudes towards premarital sex—higher education is associated with more permissive sexual values.

Do males and females have similar or different attitudes when it comes to homosexuality?

Data File: **GSS**
Task: **Auto-Analyzer**
Variable: **91) HOMO.SEX**
➤ View: **Sex**

HOMO.SEX -- What about sexual relations between two adults of the same sex -- do you think it is always wrong, almost always wrong, wrong only sometimes, or not wrong at all? (HOMOSEX)

	MALE	FEMALE
ALWAYS	59.7%	57.4%
ALMOST AL.	13.3%	11.9%
NOT WRONG	27.0%	30.7%
Number of cases	1505	1945

Females and males are about equally likely to think sex between homosexual people is always wrong.

Among females, 30.7% think sex between homosexual people is not wrong. Among males, this percentage was only 27.0%. The difference is statistically significant.

Although the auto-analyzer tells us that there is a statistically significant difference between males and females on this issue, these differences are too small to be considered substantive. It is fair to say that women and men share similar views of homosexuality.

Finally, let's look at how religion affects people's attitudes toward homosexuality.

Data File: **GSS**

Task: **Auto-Analyzer**

Variable: **91) HOMO.SEX**

➤ View: **Religion**

HOMO.SEX -- What about sexual relations between two adults of the same sex -- do you think it is always wrong, almost always wrong, wrong only sometimes, or not wrong at all? (HOMOSEX)

	LIB. PROT.	CON.PROT.	CATHOLIC	JEWISH	NONE
ALWAYS	57.5%	78.0%	51.7%	21.6%	31.4%
ALMOST AL.	14.4%	8.0%	15.4%	14.9%	15.3%
NOT WRONG	28.1%	14.0%	32.9%	63.5%	53.4%
Number of cases	694	1025	831	74	472

Conservative Protestants are most likely (78.0%) to think sex between homosexual people is always wrong and Jews (21.6%) are least likely. The difference is statistically significant.

Jews are most likely (63.5%) to think sex between homosexual people is not wrong and conservative Protestants (14.0%) are least likely. The difference is statistically significant.

The trend here is similar to attitudes toward premarital sex in that conservative Protestants are the most likely to think that homosexuality is always wrong and those who are Jewish or not religious are the least likely to hold this belief. However, the percentages of the respondents across all of the religious categories who believe homosexuality is always wrong are much higher than the percentages who believe premarital sex is always wrong. The percentages of those who believe premarital sex is always wrong range from 5.8 to 46.9 percent, whereas the percentages of those who believe homosexuality is always wrong range from a low of 21.6 percent (Jewish) to a high of 78 percent (conservative Protestant). We also see a difference in the two groups in the middle. Whereas 23.3 percent of liberal Protestants believe premarital sex is always wrong, 57.5 percent believe homosexuality is always wrong. Similarly, 18.3 percent of Catholics believe premarital sex is always wrong compared to 51.7 percent who believe homosexuality is always wrong. So, overall, religious groups tend to have stronger attitudes against homosexuality than against premarital sex in general.

When looking at attitudes toward premarital sex, we saw that the percentage of those who believe premarital sex is always wrong has declined since the 1970s. Let's see if a similar trend has occurred with regard to homosexuality.

➤ Data File: **TRENDS**

➤ Task: **Historical Trends**

➤ Variable: **30) HOMO.SEX**

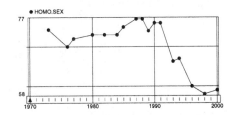

Percentage who believe homosexuality is always wrong

The percentage of respondents who believe homosexuality is always wrong was much higher in 1972 than in 2000. However, rather than being a gradual decline like we saw in attitudes toward premarital sex, the decline in disapproval of homosexuality is rather sudden. Attitudes toward homosexuality are fairly stable from 1972 to 1991, then there is a drop of nearly 10 percentage points in those who say homosexuality is always wrong between 1991 (75.5 percent) and 1993 (66.3 percent). Another large drop occurs in the 1996 survey. Whereas more permissive attitudes toward premarital sex has been a steady trend over the past few decades, more permissive attitudes toward homosexuality are rather recent.

Let's turn now to the GLOBAL data set to see how those in the United States compare to people in other countries with regard to how they answered this question.

GAY SEX: Percent who believe homosexuality is never acceptable behavior

➤ *Data File:* **GLOBAL**
➤ *Task:* **Mapping**
➤ *Variable 1:* **64) GAY SEX**
➤ *View:* **List: Rank**

RANK	CASE NAME	VALUE
1	India	95
2	China	92
3	South Korea	90
4	Russia	89
5	Lithuania	88
5	Romania	86
7	Turkey	84
8	Latvia	82
9	Belarus	80
9	Bulgaria	80

This ranking lists the proportion of the population in each country who believe that homosexuality is always wrong. The three countries where this belief is most prevalent, India (95 percent), China (92 percent), and South Korea (90 percent), are all located in Asia. Most of the remaining countries near the top of the list are located in either Eastern Europe or Asia. The Netherlands (12 percent) has the lowest percentage of citizens who believe homosexuality is always wrong. The United States is located near the middle with 53 percent.

Earlier, we saw that more permissive sexual attitudes tend to be related to higher levels of education and lower levels of religiosity. So, let's use the correlation task to see how global attitudes toward homosexuality are related to education and religion around the world. Because education is highly correlated with economic development, we will include an economic variable as well.

Data File: **GLOBAL**
➤ *Task:* **Correlation**
➤ *Variables:* **64) GAY SEX**
　　　　25) PUB EDUCAT
　　　　58) REL.PERSON
　　　　22) $ PER CAP

Correlation Coefficients
PAIRWISE deletion (1-tailed test)　　Significance Levels: ** = .01, * = .05

	GAY SEX	PUB EDUCAT	REL.PERSON	$ PER CAP
GAY SEX	1.000 (41)	-0.493 ** (39)	-0.007 (40)	-0.770 ** (41)
PUB EDUCAT	-0.493 ** (39)	1.000 (123)	-0.121 (38)	0.197 * (123)
REL.PERSON	-0.007 (40)	-0.121 (38)	1.000 (40)	0.023 (40)
$ PER CAP	-0.770 ** (41)	0.197 * (123)	0.023 (40)	1.000 (173)

Here we see that there is a strong negative correlation between GAY SEX and PUB EDUCAT—meaning that the more money a country spends on education, the lower the percentage of people who believe homosexuality is always wrong (–.493**). The negative correlation between income per capita and attitudes toward homosexuality is even stronger (–.770**). On the other hand, the correlation between attitudes toward homosexuality and church attendance is not statistically significant. Globally, it would appear that economic development and education have more of an influence an attitudes toward homosexuality than do levels of religiosity.

So far, our focus has been on people's attitudes, but the GSS also contains questions on sex-related behaviors. In the worksheets, you will have the opportunity to explore some of those variables on your own.

Marriage and Family

WORKSHEET

CHAPTER 4

NAME: _____

COURSE: _____

DATE: _____

Workbook exercises and software are copyrighted. Copying is prohibited by law.

REVIEW QUESTIONS

Based on the first part of this exercise, answer True or False to the following items:

Most Americans believe that premarital sex is always wrong, even for adults.	T F
Premarital sex was condoned in more than half of the preindustrial societies examined.	T F
Almost all societies that strictly prohibit sexual behavior of adolescent boys also prohibit sexual behavior of adolescent girls.	T F
Single adults are less likely to believe that premarital sex is wrong.	T F
Men are less likely than women to believe that premarital sex is wrong.	T F
Those who have more education tend to have more permissive sexual attitudes.	T F
People who believe premarital sex is always wrong are more likely to favor sex education in public schools.	T F
Most religious groups are just as likely to believe that premarital sex is always wrong as believe that homosexuality is always wrong.	T F

EXPLORIT QUESTIONS

1. How frequently do those who have never been married have sexual intercourse? The GSS asked respondents about their frequency of sexual intercourse in the previous 12 months.

> ➤ Data File: **GSS**
> ➤ Task: **Univariate**
> ➤ Primary Variable: **99) SEX FREQ.**
> ➤ Subset Variable: **3) EVER MAR?**
> ➤ Subset Category: **Include: 2) NO**
> ➤ View: **Pie**

a. What percentage had not had any sex?   _____%

b. What percentage had sex monthly or less?   _____%

  c.   What percentage had sex once a week or more?   _____%

2.   Is a person's attitude toward premarital sex a good predictor of his or her sexual activity?

  a.   Answer this question before doing the analysis: Of those who believe premarital sex is always wrong, what percentage do you think will have *not* had sex in the previous 12 months?   _____

Data File:	**GSS**
➤ Task:	**Cross-tabulation**
➤ Row Variable:	**99) SEX FREQ.**
➤ Column Variable:	**88) PREM.SEX**
➤ Subset Variable:	**3) EVER MAR?**
➤ Subset Category:	**Include: 2) No**
➤ View:	**Tables**
➤ Display:	**Column %**

  b.   Fill in the percentaged results for the *top* row of this table.

	ALWAYS	ALMOST AL.	NOT WRONG
NOT AT ALL	_____%	_____%	_____%

  c.   Is the percentage of those who had not had sex higher or lower than you predicted?   Higher   Lower

  d.   What is the value of V?   V = _____

  e.   Is V statistically significant?   Yes   No

  f.   Overall, is a person's attitude toward premarital sex a good predictor of his or her sexual activity?   Yes   No

3.   In the first part of this chapter, we saw that those who attend church are more likely to believe that premarital sex is always wrong. Now, test the hypothesis that **those who attend church weekly will be less sexually active than those who attend church less often.**

Data File:	**GSS**
Task:	**Cross-tabulation**
Row Variable:	**99) SEX FREQ.**
➤ Column Variable:	**50) ATTEND**
➤ Subset Variable:	**3) EVER MAR?**
➤ Subset Category:	**Include: 2) No**
➤ View:	**Tables**
➤ Display:	**Column %**

a. Fill in the percentaged results for the *top* row of the table.

	NEVER	MONTH/YRLY	WEEKLY
NOT AT ALL	_____%	_____%	_____%

b. What is the value of V?

V = _____

c. Is V statistically significant?

Yes   No

d. Is the hypothesis supported?

Yes   No

e. Those who attend church once a month most closely resemble those who (circle one)
   1. attend weekly.
   2. attend a few times a year.
   3. never attend.

4. Because people with more education have more permissive sexual attitudes, **people with a higher level of education will be more sexually active.**

> *Data File:* **GSS**
> *Task:* **Cross-tabulation**
> *Row Variable:* **99) SEX FREQ.**
> ➤ *Column Variable:* **27) EDUCATION**
> ➤ *Subset Variable:* **3) EVER MAR?**
> ➤ *Subset Category:* **Include: 2) No**
> ➤ *View:* **Tables**
> ➤ *Display:* **Column %**

a. What is the value of V?

V = _____

b. Is V statistically significant?

Yes   No

c. Is the hypothesis supported?

Yes   No

d. In the first part of this chapter, it was demonstrated that people with more education tend to have more permissive sexual attitudes. How would you explain the relationship between those findings and the results of this hypothesis?

5. Earlier we saw that men tend to have more permissive sexual attitudes. So, the hypothesis is: **Single men will be more sexually active than single women**.

> Data File: **GSS**
> Task: **Cross-tabulation**
> Row Variable: **99) SEX FREQ.**
> ➤ Column Variable: **25) SEX**
> ➤ Subset Variable: **3) EVER MAR?**
> ➤ Subset Category: **Include: 2) No**
> ➤ View: **Tables**
> ➤ Display: **Column %**

a. Fill in the percentaged results for this table.

	MALE	FEMALE
NOT AT ALL	_____ %	_____ %
MONTH/LESS	_____ %	_____ %
WKLY/MORE	_____ %	_____ %

b. What is the value of V? V = _____

c. Is V statistically significant? Yes No

d. Is the hypothesis supported? Yes No

6. The GSS also asks questions about the relationship between the respondent and those with whom he or she had sex. For example, respondents were asked: "The last time you had sex, was it with someone you were in an ongoing relationship with?" Did men and women answer this question the same way?

> Data File: **GSS**
> Task: **Cross-tabulation**
> ➤ Row Variable: **103) RELATE SEX**
> ➤ Column Variable: **25) SEX**
> ➤ Subset Variable: **3) EVER MAR?**
> ➤ Subset Category: **Include: 2) No**
> ➤ View: **Tables**
> ➤ Display: **Column %**

a. What percentage of men say they are in a close relationship? _____ %

b. What percentage of women say they are in a close relationship? _____ %

c. What is the value of V? V = _____

d. Is V statistically significant?   Yes   No

e. Do you think men and women are involved in different types of relationships, or do men and women just perceive relationships differently? Explain your answer.

7. The GSS does not ask whether respondents are heterosexual or homosexual, but it does ask whether their sexual activity in the previous year has been with males, females, or both. Complete the following analysis to see how people in the entire sample responded to this question.

> Data File: **GSS**
> Task: **Cross-tabulation**
> ➤ Row Variable: **98) SEX OF SEX**
> ➤ Column Variable: **25) SEX**
> ➤ View: **Tables**
> ➤ Display: **Column %**

Remember to delete the subset variable.

a. What percentage of men had sex only with men?   _____%

b. What percentage of men had sex with both men and women?   _____%

c. What percentage of men had sex only with women?   _____%

d. What percentage of women had sex only with men?   _____%

e. What percentage of women had sex with both men and women?   _____%

f. What percentage of women had sex only with women?   _____%

Chapter 4: Premarital Sex: Attitudes and Behavior

8. Finally, because those who have regular sex are more likely to expect to have sexual intercourse, test the hypothesis that **those who have sex weekly will be the most likely to have used a condom the last time they had sex.**

> Data File: **GSS**
> Task: **Cross-tabulation**
> ➤ Row Variable: **102) CONDOM**
> ➤ Column Variable: **99) SEX FREQ.**
> ➤ Subset Variable: **3) EVER MAR?**
> ➤ Subset Category: **Include: 2) No**
> ➤ View: **Tables**
> ➤ Display: **Column %**

a. What is the value of V?

V = _____

b. Is V statistically significant?

Yes   No

c. Which group was the least likely to have used a condom? (circle one)
   1. those who had sex weekly the previous 12 months
   2. those who had sex monthly or less the previous 12 months
   3. those who had not had sex in the previous 12 months

d. Is the hypothesis supported?

Yes   No

e. How would you explain these results?

CHAPTER 5

MARITAL AND EXTRAMARITAL SEX

Tasks: Univariate, Cross-tabulation, Historical Trends, Mapping
Data Files: GSS, TRENDS, GLOBAL, CULTURES

Sexual intercourse is an aspect of married life that most people value but are often reluctant to talk about. As a result, there is some uncertainty as to what is *typical* with regard to a married couple's sex life. For example, how often do most couples have sexual intercourse? How does the presence of children in the household affect a couple's sex life? How common are extramarital affairs? These are a few of the questions we will try to address in this chapter.

Let's begin by looking at how often married individuals say they have sexual intercourse (Note that the subset option must be used in order to limit the analysis to individuals who are married.)

➤ *Data File:* **GSS**
➤ *Task:* **Univariate**
➤ *Primary Variable:* **99) SEX FREQ.**
➤ *Subset Variable:* **2) MARRIED?**
➤ *Subset Category:* **Include: 1) Yes**
➤ *View:* **Pie**

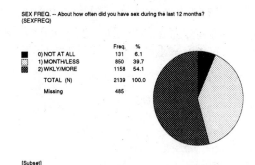

SEX FREQ. -- About how often did you have sex during the last 12 months? (SEXFREQ)

		Freq.	%
■	0) NOT AT ALL	131	6.1
▨	1) MONTH/LESS	850	39.7
▨	2) WKLY/MORE	1158	54.1
	TOTAL (N)	2139	100.0
	Missing	485	

[Subset]

As one might expect, there is some variation in the frequency of sexual intercourse among married couples. However, over half (54.1 percent) of the married respondents said they have sexual intercourse at least once per week. About 40 percent fall into the middle range of monthly or less frequently. The smallest category is those who say they never had sexual intercourse in the past 12 months (6.1 percent). Let's see if we can uncover some of the factors that predict how sexually active a married couple is likely to be.

We will begin by looking at the relationship between marital happiness and the frequency of sexual intercourse. Are those who are more happily married more likely to have sex at least once a week?

Data File: **GSS**
➤ Task: **Cross-tabulation**
➤ Row Variable: **99) SEX FREQ.**
➤ Column Variable: **74) HAP.MARR.?**
➤ View: **Tables**
➤ Display: **Column %**

SEX FREQ. by HAP.MARR.?
Cramer's V: 0.104 **

		HAP.MARR.?				
		VERY HAPPY	PRET.HAPPY	NOT TOO	Missing	TOTAL
SEX FREQ	NOT AT ALL	76	46	9	894	131
		5.6%	6.6%	14.1%		6.2%
	MONTH/LESS	488	315	39	720	842
		35.7%	45.3%	60.9%		39.6%
	WKLY/MORE	803	334	16	844	1153
		58.7%	48.1%	25.0%		54.2%
	Missing	277	184	14	590	1065
	TOTAL	1367	695	64	3048	2126
		100.0%	100.0%	100.0%		

It appears that there is a relationship between marital happiness and the likelihood that a married couple has sex at least weekly. Among the very happily married, 58.7 percent have sexual intercourse at least once a week, as compared to 48.1 percent for those who are pretty happy, and 25 percent for those who are not too happily married. The difference is statistically significant (V = .104**). Of course, we cannot tell which is the cause and which is the effect in this relationship. Do people have sexual intercourse more frequently because they are happily married, or are people more happily married because they have sex frequently? Most likely, marital happiness and the frequency of sex are very interrelated. Those who are happily married probably have more frequent sexual intercourse, which, in turn increases their level of marital happiness.

For health-related reasons, one would think that **older couples will be less likely to have sexual intercourse weekly**.

Data File: **GSS**
Task: **Cross-tabulation**
Row Variable: **99) SEX FREQ.**
➤ Column Variable: **24) AGE**
➤ Subset Variable: **2) MARRIED?**
➤ Subset Category: **Include: 1) Yes**
➤ View: **Tables**
➤ Display: **Column %**

SEX FREQ. by AGE
Cramer's V: 0.279 **

		AGE				
		<30	30-49	50 AND UP	Missing	TOTAL
SEX FREQ	NOT AT ALL	2	6	123	0	131
		0.9%	0.5%	16.0%		6.1%
	MONTH/LESS	44	425	381	0	850
		19.6%	37.1%	49.7%		39.8%
	WKLY/MORE	178	716	263	1	1157
		79.5%	62.4%	34.3%		54.1%
	Missing	35	206	240	4	485
	TOTAL	224	1147	767	5	2138
		100.0%	100.0%	100.0%		

Once again, select 2) MARRIED? as your subset variable. Include 1) Yes as your subset category.

The direction of the results is clearly consistent with the hypothesis. Married people under age 30 are the most likely to have sexual intercourse at least once a week (79.5 percent). The next most sexually active group is the 30- to 49-year-olds, 62.4 percent of whom have sexual intercourse weekly. Among those who are 50 or older, only 34.3 percent have sexual intercourse once a week or more. The top row shows that only 16 percent of the oldest married respondents have not had sexual intercourse in the last 12 months. Most of these respondents had sex monthly or less. There is a statistically significant relationship between the frequency of sexual intercourse and age (V = .279**), so the hypothesis is supported.

Another factor that may influence the frequency of sexual intercourse is the presence of children. Children in the home can limit privacy and opportunities for spontaneity, and also drain emotional reserves and romantic energy. Thus, the hypothesis is: **Couples who have children living at home will be less likely to have sexual intercourse weekly**. We have already seen that those over age 50 are far less likely to have sex weekly, so we will use 11) FAM.STAGE to separate those who have never had children from those with grown children who are in the empty-nest category.

Data File: **GSS**
Task: **Cross-tabulation**
Row Variable: **99) SEX FREQ.**
➤ Column Variable: **11) FAM.STAGE**
➤ View: **Tables**
➤ Display: **Column %**

SEX FREQ.   by   FAM.STAGE
Cramer's V: 0.233 **

		FAM.STAGE				
		NO KIDS	WITH KIDS	EMPTY NEST	Missing	TOTAL
SEX FREQ.	NOT AT ALL	1	14	107	903	122
		0.5%	1.3%	14.2%		6.0%
	MONTH/LESS	63	377	360	762	800
		29.2%	35.2%	47.7%		39.2%
	WKLY/MORE	152	679	288	878	1119
		70.4%	63.5%	38.1%		54.8%
	Missing	29	187	242	607	1065
	TOTAL	216	1070	755	3150	2041
		100.0%	100.0%	100.0%		

The subset category for marital status is not necessary because 11) FAM.STAGE includes only married respondents.

There is a statistically significant relationship here (V = .233**), but the dividing point is clearly between the empty-nest category (38.1 percent) and the other two categories. Among married couples with no children, 70.4 percent have sexual intercourse at least once a week; this is only seven percentage points greater than those who have children living at home (63.5 percent). So, what happens if we exclude the empty-nest category from the analysis? Will there still be a statistically significant relationship between family stage and the frequency of sexual intercourse?

Data File: **GSS**
Task: **Cross-tabulation**
Row Variable: **99) SEX FREQ.**
➤ Column Variable: **11) FAM.STAGE**
➤ Subset Variable: **11) FAM.STAGE**
➤ Subset Category: **Exclude: 1) EMPTY NEST**
➤ View: **Tables**
➤ Display: **Column %**

SEX FREQ.   by   FAM.STAGE

Cramer's V: 0.058

		FAM.STAGE		
		NO KIDS	WITH KIDS	TOTAL
SEX FREQ.	NOT AT ALL	1	14	15
		0.5%	1.3%	1.2%
	MONTH/LESS	63	377	440
		29.2%	35.2%	34.2%
	WKLY/MORE	152	679	831
		70.4%	63.5%	64.6%
	Missing	29	187	216
	TOTAL	216	1070	1286
		100.0%	100.0%	

Excluding the empty-nest category does not change the percentages in the other two columns, but it does change the value of V (V = .058) to the point that it is no longer statistically significant. Therefore, the hypothesis about having children and the frequency of sexual intercourse is rejected. Perhaps if we were able to analyze the actual frequency of sexual intercourse rather than using these broad categories we might find a difference. But when it comes to having sexual intercourse at least weekly, there is not a statistically significant relationship between sex and having children living at home.

Let's look at one more factor that may influence the frequency of sexual intercourse for married couples: religion. Conservative religious groups have traditionally had very clear proscriptions against premarital sex. Does this have a lingering effect on the frequency of sex after marriage? The hypothesis is: **Couples who attend fundamentalist churches will be less likely to have sex weekly.**

Data File:	**GSS**
Task:	**Cross-tabulation**
Row Variable:	**99) SEX FREQ.**
➤ Column Variable:	**49) R.FUND/LIB**
➤ Subset Variable:	**2) MARRIED?**
➤ Subset Category:	**Include: 1) Yes**
➤ View:	**Tables**
➤ Display:	**Column %**

SEX FREQ.   by   R.FUND/LIB

Cramer's V: 0.057 *

		R.FUND/LIB				
		FUNDAMENT.	MODERATE	LIBERAL	Missing	TOTAL
SEX FREQ.	NOT AT ALL	46	48	34	3	128
		7.3%	6.1%	5.9%		6.4%
	MONTH/LESS	228	303	263	56	794
		36.1%	38.6%	45.7%		39.9%
	WKLY/MORE	358	433	279	88	1070
		56.6%	55.2%	48.4%		53.7%
	Missing	142	193	119	31	485
	TOTAL	632	784	576	178	1992
		100.0%	100.0%	100.0%		

According to these results, virtually the same percentage of religious fundamentalists (56.6 percent) and moderates (55.2 percent) have sexual intercourse at least weekly. Religious liberals are somewhat less likely to have weekly sex (48.4 percent), but the difference is not great. The hypothesis is rejected; religious fundamentalists are not less likely to have sexual intercourse at least once a week.

Is the United States different from other nations in terms of attitudes toward marital sex? The following question was asked in over 40 countries as part of the World Values Survey: "Do you and your (spouse) share sexual attitudes?" Let's look at the percentage of people in each nation who say they do share sexual attitudes with their spouse.

➤ Data File:	**GLOBAL**
➤ Task:	**Mapping**
➤ Variable 1:	**42) SPOUSE SEX**
➤ View:	**List: Rank**

SPOUSE SEX: Percent who say they and their spouse share sexual attitudes

RANK	CASE NAME	VALUE
1	Iceland	71
2	Norway	70
3	Sweden	68
4	Turkey	67
4	Switzerland	67
6	Denmark	66
7	United States	65
7	Netherlands	65
9	Hungary	64
10	Argentina	63

People in Iceland (71 percent), Norway (70 percent), and Sweden (68 percent) are the most likely to share sexual attitudes with their spouse. A very different situation is found in Japan (21 percent), China (31 percent), and Lithuania (38 percent), where only about a third of people share sexual attitudes with their spouse. The United States ranks toward the top of the list with 65 percent.

Do people across the world differ in whether they think a happy sexual relationship is very important for a happy marriage?

HAPPY SEX?: Percent who think a happy sexual relationship is very important for a happy marriage

Data File: **GLOBAL**
Task: **Mapping**
➤ Variable 1: **43) HAPPY SEX?**
➤ View: **List: Rank**

RANK	CASE NAME	VALUE
1	Nigeria	93
2	Argentina	81
3	Mexico	77
4	Chile	74
5	Turkey	73
5	India	73
5	Iceland	73
8	Brazil	72
8	Canada	72
10	Italy	70

People in Nigeria (93 percent), Argentina (81 percent), and Mexico (77 percent) are the most likely to believe that a happy sexual relationship is very important for a happy marriage. The United States ranks eleventh at 69 percent. Once again at the bottom of the list we find China (39 percent), Lithuania (37 percent), and Japan (28 percent). Although we won't examine the relationship here, the similarity between these two ranked lists suggests that these variables are highly correlated.

So far we have looked only at the issue of sexual intercourse between marital partners. What attitudes do people throughout the world have toward extramarital sexual intercourse?

EX-MARITAL: Percent who think it is never OK for married people to have an affair

Data File: **GLOBAL**
Task: **Mapping**
➤ Variable 1: **62) EX-MARITAL**
➤ View: **List: Rank**

RANK	CASE NAME	VALUE
1	India	91
2	Turkey	87
3	China	72
4	South Korea	71
4	Iceland	71
6	United States	70
6	South Africa	70
6	Argentina	70
9	Ireland	69
10	Poland	67

Here we see the percent who think it is *never* OK for married people to have an affair. An astounding range is again evident. India (91 percent) comes in with the highest ranking, followed by Turkey (87 percent), and China (72 percent). Only 26 percent of people in the Czech Republic and 30 percent of people in Estonia think it is never OK to have an affair. Back toward the top of the list we see that over two-thirds of people in the United States (70 percent) think it is never OK to have an affair.

What about prostitution?

PROSTITUTE: Percent who think that prostitution is never OK

Data File: **GLOBAL**
Task: **Mapping**
➤ Variable 1: **66) PROSTITUTE**
➤ View: **List: Rank**

RANK	CASE NAME	VALUE
1	China	92
2	India	84
3	Turkey	82
4	Lithuania	78
4	South Korea	78
4	Romania	78
7	Russia	77
7	Chile	77
9	Poland	75
10	Brazil	73

Of those who think prostitution is never acceptable, China (92 percent), India (84 percent), and other Eastern nations dominate the top of the list. Western European countries firmly anchor the bottom of the list, with only 1 in 5 persons in the Netherlands (19 percent) and about 1 in 3 persons in Finland (29 percent) believing that prostitution is never acceptable.

So now we have looked at the frequency of sexual intercourse between marital partners in the United States as well as attitudes toward marital and extramarital sexual intercourse across nations of the world. In the worksheet section that follows, we will continue our examination of extramarital sex by looking at the sexual standards of preindustrial societies and by exploring the nature and frequency of extramarital affairs in the United States.

WORKSHEET

CHAPTER 5

NAME:

COURSE:

DATE:

REVIEW QUESTIONS

Based on the first part of this chapter, answer True or False to the following items:

When people in the United States were asked how frequently they had sexual intercourse, the most common response was at least once a week.	T	F
Those who say they are very happily married have sexual intercourse more frequently than those who are less happily married.	T	F
Contrary to popular opinion, the frequency of sexual intercourse does not decline significantly with age for those who are married.	T	F
There is a statistically significant difference in the frequency of sexual intercourse for married couples who have children living at home in comparison to married couples without children at home.	T	F
Married religious fundamentalists are less likely to have sex at least weekly.	T	F
Lithuania, China, and Japan are very similar in terms of the attitudes people have regarding a happy sexual relationship being related to a happy marriage. They are also similar in the sense that people in all three nations do not share sexual attitudes with their spouses.	T	F
People in Western Europe are the least likely to think that prostitution is never OK.	T	F

EXPLORIT QUESTIONS

1. Before we look at the issue of extramarital sex in modern times, let's briefly examine its frequency in preindustrial societies. The variable 65) FEM EXTRA measures whether or not married women can have extramarital affairs, and if they do.

    ➤ Data File: **CULTURES**
    ➤ Task: **Univariate**
    ➤ Primary Variable: **65) FEM EXTRA**
    ➤ View: **Pie**

    a. In what percentage of preindustrial societies did women abstain from extramarital affairs? _____%

b. In what percentage of preindustrial societies were women permitted to have extramarital affairs?   _____%

2. In the United States it has often been argued that a double standard exists in terms of sexual behavior —a standard that places more restrictions on women than men. Did double standards for extramarital sex exist in preindustrial societies?

> *Data File:* **CULTURES**
> *Task:* **Univariate**
> ➤ *Primary Variable:* **64) EXTRAMARTL**
> ➤ *View:* **Pie**

a. Did a double standard exist in the majority of preindustrial societies?   Yes   No

b. In what percentage of preindustrial societies did the double standard favor males?   _____%

3. In the first part of this chapter, we used data from the World Values Survey to determine that around 70 percent of people in the United States believe that extramarital affairs are always wrong. Let's examine a similar survey question that was asked in the General Social Survey.

> ➤ *Data File:* **GSS**
> ➤ *Task:* **Univariate**
> ➤ *Primary Variable:* **90) XMAR.SEX**
> ➤ *View:* **Pie**

a. What percentage of the U.S. population believes extramarital sex is always wrong?   _____%

b. What percentage of the population believes extramarital sex is not wrong?   _____%

c. Are these results similar to those found in the World Values Survey?   Yes   No

4. Construct the following graph to see how attitudes toward extramarital sex changed from 1972 to 2000.

> ➤ *Data File:* **TRENDS**
> ➤ *Task:* **Historical Trends**
> ➤ *Variable:* **21) XMAR.SEX%**

Each point on the graph illustrates the percentage of the population who believe extramarital sex is always wrong.

a. The percentage of people who believe extramarital sex is always wrong has declined since 1972.   T   F

b.   Are these the results you expected? What might explain this trend line?

5.   For this question, we will return to the GSS file and look at how frequently extramarital affairs actually occur. However, before you do the analysis, I want you to guess what percentage of the U.S. population (those who have ever been married) will say they have had an extramarital affair.

a.   What is your guess?   _____%

> ➤ *Data File:* **GSS**
> ➤ *Task:* **Univariate**
> ➤ *Primary Variable:* **101) EVER STRAY**
> ➤ *View:* **Pie**

b.   What percentage of the population say they have ever had an extramarital affair?   _____%

c.   Is this percentage higher or lower than you expected?   Higher   Lower

d.   How would you explain the difference between your prediction and the actual results?

6.   Now, let's look at the relationship between attitudes toward extramarital sex and whether or not people say they have ever had an affair. We would expect to find that **people who believe an extramarital affair is always wrong will be less likely to have had an affair.**

> *Data File:* **GSS**
> ➤ *Task:* **Cross-tabulation**
> ➤ *Row Variable:* **101) EVER STRAY**
> ➤ *Column Variable:* **90) XMAR.SEX**
> ➤ *View:* **Tables**
> ➤ *Display:* **Column %**

a. Fill in the percentaged results for the *top* row of the table.

	ALWAYS	ALMOST AL.	NOT WRONG
YES	_____%	_____%	_____%

b. What is the value of Cramer's V?   V = _____

c. Is V statistically significant?   Yes   No

d. People who believe extramarital affairs are *almost always wrong* most closely
   resemble those who believe extramarital affairs are *not wrong*.   T   F

e. Most people who believe extramarital affairs are *not wrong* have had an affair.   T   F

f. The more people disapprove of extramarital affairs, the less likely they are to have one.   T   F

7. Although the actual number of people in this survey who admit to having had an extramarital affair is
   small, let's take a look at people who say they have had sexual intercourse with (other than their
   spouse) in the past year. (Note: The next set of survey questions include only those people who indicate they had sex with someone other than their spouse in the past year. We are further limiting this
   analysis to those who are currently married. Thus, our results will include only those married individuals who have had an affair in the past year.)

> Data File: **GSS**
> ➤ Task: **Univariate**
> ➤ Primary Variable: **95) ACQNT SEX?**
> ➤ Subset Variable: **2) MARRIED?**
> ➤ Subset Category: **Include: 1) Yes**
> ➤ View: **Pie**

What percentage of the married respondents who had an extramarital affair in the
previous year had sexual intercourse with an acquaintance?   _____%

8. Did married individuals who had an extramarital affair in the previous year have sexual intercourse
   with a close friend?

> Data File: **GSS**
> Task: **Univariate**
> ➤ Primary Variable: **94) FRIEND SEX**
> ➤ Subset Variable: **2) MARRIED?**
> ➤ Subset Category: **Include: 1) Yes**
> ➤ View: **Pie**

What percentage of the married respondents who had an extramarital affair in the
previous year had sexual intercourse with a close friend?   _____%

9. Did married individuals who had an extramarital affair in the previous year have sexual intercourse with a casual date or someone they picked up?

> Data File: **GSS**
> Task: **Univariate**
> ➤ Primary Variable: **96) PIKUP SEX?**
> ➤ Subset Variable: **2) MARRIED?**
> ➤ Subset Category: **Include: 1) Yes**
> ➤ View: **Pie**

What percentage of the married respondents who had an extramarital affair in the previous year had sexual intercourse with a casual date or someone they picked up?   _____%

10. How about with someone they paid to have sexual intercourse?

> Data File: **GSS**
> Task: **Univariate**
> ➤ Primary Variable: **97) PAID SEX?**
> ➤ Subset Variable: **2)MARRIED?**
> ➤ Subset Category: **Include: 1) Yes**
> ➤ View: **Pie**

a. What percentage of the married respondents who had an extramarital affair in the previous year had sexual intercourse with someone they paid to have sex?   _____%

b. Based on the results for Questions 7–10, most people who have had an extramarital affair in the previous year did so with (circle one)
   1. an acquaintance.
   2. a friend.
   3. someone they had just met.
   4. someone they paid for sex.

c. People are the least likely to have an extramarital affair with (circle one)
   1. an acquaintance.
   2. a friend.
   3. someone they had just met.
   4. someone they paid for sex.

d. What do the results of Questions 7–10 suggest to you about the nature of extramarital sexual activity?

11. The hypothesis is: **People who describe their marriage as very happy will be less likely to have ever had an extramarital affair.**

> Data File: **GSS**
> ➤ Task: **Cross-tabulation**
> ➤ Row Variable: **101) EVER STRAY**
> ➤ Column Variable: **74) HAP.MARR.?**
> ➤ View: **Tables**
> ➤ Display: **Column %**

a. Fill in the percentaged results for the *top* row of the table.

	VERY HAPPY	PRET. HAPPY	NOT TOO
YES	_____%	_____%	_____%

b. What is the value of Cramer's V? $V =$ _____

c. Is V statistically significant?   Yes   No

d. Is the hypothesis supported?   Yes   No

12. Let's return to our analysis of marital sex. Whereas in the past, X-rated movies were shown only in adult theaters, today, sex-related videos are widely accessible in the United States. Let's see what percentage of married respondents have viewed an X-rated movie in the past year.

a. Before conducting your analysis, what percentage of those who are currently married would you guess have watched an X-rated in the previous 12 months?   _____%

> Data File: **GSS**
> ➤ Task: **Univariate**
> ➤ Primary Variable: **92) X-MOVIE?**
> ➤ Subset Variable: **2) MARRIED?**
> ➤ Subset Category: **Include: 2) Yes**
> ➤ View: **Pie**

b. What percentage of married respondents have watched an X-rated movie? _____%

13. Are married men more likely than married women to have watched an X-rated movie?

>
> Data File: **GSS**
> ➤ Task: **Cross-tabulation**
> ➤ Row Variable: **92) X-MOVIE**
> ➤ Column Variable: **25) SEX**
> ➤ Subset Variable: **2) MARRIED?**
> ➤ Subset Category: **Include: 2) Yes**
> ➤ View: **Tables**
> ➤ Display: **Column %**

a. Fill in the percentaged results for the *top* row of the table.

	MALE	FEMALE
YES	_____%	_____%

b. Based on these results, do you think it is likely that most married respondents who watched an X-rated movie did so with their spouse? Explain your answer.

CHAPTER **6**

SINGLEHOOD

Tasks: Univariate, Historical Trends, Cross-tabulation, Auto-Analyzer
Data Files: GSS, TRENDS

The status of being *single* is one of the most diverse statuses in our society. As with all marital statuses, singles differ by age, race, social class, and so on, but singles also differ in the way they have acquired their status of being single. The most traditional profile of a single adult may be a female or male in their early twenties who is waiting for the right time or person to marry. But there are other types of singles as well. For example, there are those who plan to be single their whole life, there are singles with children and without, and there are singles who are heterosexual and those who are homosexual. And there are also those who have been married at some time, but who are now single as the result of divorce or the death of a spouse. All of these groups can be included in the status of being *single*. But, because other chapters in this book include analyses for all those who are unmarried (Chapter 8), and those who are currently divorced (Chapter 13), the analyses in this chapter will be limited to those who have never been married.

What does it mean to be *single* in our society? What affect does being single have on a person's lifestyle? Do singles engage in different activities than those who are married? Do singles find life to be more or less exciting? The goal of this chapter is to explore the answers to these questions by comparing the lifestyles of singles with the lifestyles of those who are currently married.

Let's start by looking at the percentage of General Social Survey respondents who indicate that they have never been married.

➤ *Data File:* **GSS**
➤ *Task:* **Univariate**
➤ *Primary Variable:* **3) EVER MAR?**
➤ *View:* **Pie**

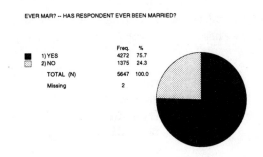

EVER MAR? -- HAS RESPONDENT EVER BEEN MARRIED?

		Freq.	%
■	1) YES	4272	75.7
▧	2) NO	1375	24.3
	TOTAL (N)	5647	100.0
	Missing	2	

According to these results, 24.3 percent of the respondents have never been married. How does this percentage compare with the results in earlier years of the GSS?

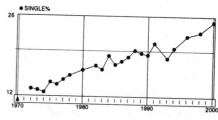

> *Data File:* **TRENDS**
> > *Task:* **Historical Trends**
> > *Variable:* **6) SINGLE%**

Percentage of GSS respondents who are single

This graph illustrates the percentage of respondents who indicated they had never been married from 1972 to 2000. As you can see, the size of the singles population in the GSS sample has doubled during this relatively brief period of time. Why has the percentage of singles grown so dramatically? Part of the explanation can be found by looking at trends in education during the same period of time.

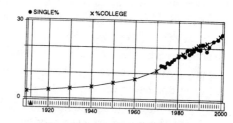

> *Data File:* **TRENDS**
> *Task:* **Historical Trends**
> > *Variables:* **6) SINGLE%**
> **25) %COLLEGE**

Percentage of GSS respondents who are single
Percentage of adults with college education

This graph plots both the percentage of singles and the percentage of college graduates. The data on college graduates starts with 1910 and goes to 1999, whereas the data on singles ranges from 1972 to 2000. But, if you look at the years where both data are available, you will see that the two trends closely parallel each other. The percentage of singles rises at the same rate as the percentage of college graduates. It could be that people are delaying marriage to go to college. Or perhaps college graduates are more likely to forego marriage altogether to pursue a career. Either way, one reason for the decline in the percentage of married couples is the increase in the percentage of adults who have never married.

Going to college does tend to delay marriage for most people, so let's see if the rise in college graduation rates has coincided with an increase in the age at marriage for males and females.

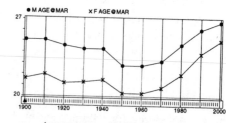

> *Data File:* **TRENDS**
> *Task:* **Historical Trends**
> > *Variables:* **12) M AGE@MAR**
> **13) F AGE@MAR**

Average age at marriage for males and females

The age at marriage for both females and males was higher in 2000 than at any other point in the past one hundred years. There was a decline in the age at marriage following WW II, but since then, the age

98

Marriage and Family

at which both men and women marry has risen steadily thereby leading to an increase in the size of the singles population.

Does this trend represent a growing disillusionment with marriage, or are these singles simply delaying marriage until time and circumstances are right? Perhaps they don't believe they have met the right person yet, but hope to get married when they do. Let's take a look at what percentage of never-married respondents are involved in a steady relationship.

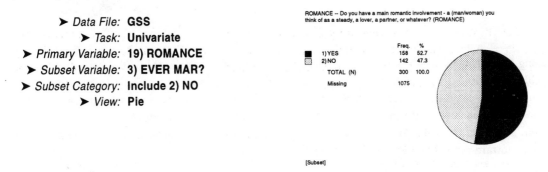

➤ *Data File:* **GSS**
➤ *Task:* **Univariate**
➤ *Primary Variable:* **19) ROMANCE**
➤ *Subset Variable:* **3) EVER MAR?**
➤ *Subset Category:* **Include 2) NO**
➤ *View:* **Pie**

By using the subset feature, we are able to limit this graph to only those who have never been married. What we find is that the singles population is split cleanly down the middle with regard to this variable; 52.7 percent of the singles are involved in a steady relationship and 47.3 percent are not. Among those who are involved in a steady relationship, how many think they will eventually marry the person they are currently dating?

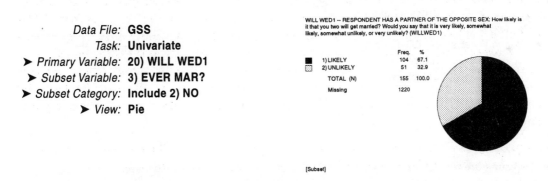

Data File: **GSS**
Task: **Univariate**
➤ *Primary Variable:* **20) WILL WED1**
➤ *Subset Variable:* **3) EVER MAR?**
➤ *Subset Category:* **Include 2) NO**
➤ *View:* **Pie**

This graph includes only singles who are involved in a steady relationship. What we find is that 67.1 percent of the respondents say they are likely to marry the person they are currently dating. For this group, their single status would appear to represent a timing issue rather than a disillusionment with marriage.

What about those who do not have a steady partner or relationship—are they planning to get married if they meet the right person?

Data File: **GSS**
Task: **Univariate**
➤ Primary Variable: **21) WILL WED2**
➤ Subset Variable: **3) EVER MAR?**
➤ Subset Category: **Include 2) NO**
➤ View: **Pie**

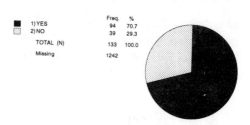

WILL WED2 -- RESPONDENT DOES NOT HAVE A STEADY LOVER OR PARTNER: If the right person came along, would you like to be married? (WILLWED2)

		Freq.	%
■	1) YES	94	70.7
▨	2) NO	39	29.3
	TOTAL (N)	133	100.0
	Missing	1242	

[Subset]

Among those singles who are not currently in a serious dating relationship, 70.7 percent say they would get married if the right person came along. Thus, it would appear that even though the size of the unmarried population is larger than it has been in the past, this trend does not represent a widespread disillusionment with marriage. Most singles are planning to marry when the time and the person are right. However, there is a sizable portion of the singles population who do not seem to be inclined toward marriage at all.

Now let's look a little more at the singles lifestyle, beginning with the type of communities in which they are likely to live. In the first chapter of this book, we found that states with higher urban populations tend to have a larger proportion of single adults. Let's see if a similar trend is evident in the General Social Survey.

Data File: **GSS**
➤ Task: **Cross-tabulation**
➤ Row Variable: **3) EVER MAR?**
➤ Column Variable: **38) COMMUNITY**
➤ View: **Tables**
➤ Display: **Column %**

EVER MAR?   by   COMMUNITY
Cramer's V: 0.166 **

		COMMUNITY					
		BIG CITY	SUBURB	SMALL TOWN	RURAL	Missing	TOTAL
EVER MAR?	YES	156	221	406	118	3371	901
		65.5%	75.7%	73.8%	92.9%		74.6%
	NO	82	71	144	9	1069	306
		34.5%	24.3%	26.2%	7.1%		25.4%
	Missing	0	0	0	0	2	2
	TOTAL	238	292	550	127	4442	1207
		100.0%	100.0%	100.0%	100.0%		

These results are consistent with the state-level analysis; the more urban an area in which a respondent lives, the more likely he or she is to be single. The biggest difference is between the 34.5 percent of big-city residents who have never married and the 7.1 percent of rural residents. Those who live in suburban neighborhoods (24.3 percent) and those who live in small towns (26.2 percent) also are more likely to be single than those who live in rural areas. While some of these singles were born and raised in the city where career and entertainment opportunities abound, others were undoubtedly drawn from less urban areas to seek those same opportunities. Either way, there is an affinity between singleness and urban lifestyles.

Let's see if, in fact, those who are single are more likely to take advantage of the entertainment opportunities in their communities.

Data File: **GSS**

Task: **Cross-tabulation**

➤ Row Variable: **85) EAT OUT**

➤ Column Variable: **7) SINGLE/MAR**

➤ View: **Tables**

➤ Display: **Column %**

EAT OUT   by   SINGLE/MAR
Cramer's V: 0.093 **

		SINGLE/MAR			
		SINGLE	MARRIED	Missing	TOTAL
EAT OUT	YES	311	550	297	861
		88.4%	81.1%		83.6%
	NO	41	128	113	169
		11.6%	18.9%		16.4%
	Missing	1023	1946	1240	4209
	TOTAL	352	678	1650	1030
		100.0%	100.0%		

The GSS asked whether respondents had eaten out in the previous seven days. Because the vast majority of Americans do eat out at least once a week, there is not a lot of variation in this variable. But, based on these results, single respondents (88.4 percent) are somewhat more likely than married respondents (81.1 percent) to have eaten at a restaurant the previous week. The difference is statistically significant (V = .093**).

How about going to a movie? Are singles more likely to have gone to see a film at a theater in the previous week?

Data File: **GSS**

Task: **Cross-tabulation**

➤ Row Variable: **86) SEE FILM**

➤ Column Variable: **7) SINGLE/MAR**

➤ View: **Tables**

➤ Display: **Column %**

SEE FILM   by   SINGLE/MAR
Cramer's V: 0.132 **

		SINGLE/MAR			
		SINGLE	MARRIED	Missing	TOTAL
SEE FILM	YES	76	79	38	155
		21.6%	11.6%		15.0%
	NO	276	600	371	876
		78.4%	88.4%		85.0%
	Missing	1023	1945	1241	4209
	TOTAL	352	679	1650	1031
		100.0%	100.0%		

Again, there is a statistically significant difference between those who are single and those who are married (V = .132**). Those who are single (21.6 percent) are almost twice as likely as those who are married (11.6 percent) to have gone to see a movie the previous week.

Next, let's look at concert attendance. This time the GSS asks respondents whether they have gone to a pop music concert (rock, country, rap, etc.) in the previous 12 months.

	Data File:	**GSS**
	Task:	**Cross-tabulation**
➤	Row Variable:	**84) POP MUSIC**
➤	Column Variable:	**7) SINGLE/MAR**
➤	View:	**Tables**
➤	Display:	**Column %**

POP MUSIC   by   SINGLE/MAR
Cramer's V: 0.160 **

		SINGLE/MAR			
		SINGLE	MARRIED	Missing	TOTAL
POP MUSIC	YES	183	241	127	424
		52.1%	35.5%		41.2%
	NO	168	437	279	605
		47.9%	64.5%		58.8%
	Missing	1024	1946	1244	4214
	TOTAL	351	678	1650	1029
		100.0%	100.0%		

The difference between those who are single and those who are married is larger in this table. More than half of the single respondents, 52.1 percent, have attended a concert compared to 35.5 percent of those who are married. The difference is statistically significant (V = .160**). But could this just be an age difference? After all, those who are single tend to be younger than those are married. Maybe the difference here is age and not marital status. Let's see what happens if we exclude the oldest respondents from the analysis and compare only those under age 50. Are those who are single still more likely to have attended a concert?

	Data File:	**GSS**
	Task:	**Cross-tabulation**
	Row Variable:	**84) POP MUSIC**
	Column Variable:	**7) SINGLE/MAR**
➤	Subset Variable:	**24) AGE**
➤	Subset Category:	**Exclude: 3) 50 and up**
➤	View:	**Tables**
➤	Display:	**Column %**

POP MUSIC   by   SINGLE/MAR

Cramer's V: 0.147 **

		SINGLE/MAR			
		SINGLE	MARRIED	Missing	TOTAL
POP MUSIC	YES	176	161	77	337
		54.3%	39.6%		46.1%
	NO	148	246	101	394
		45.7%	60.4%		53.9%
	Missing	904	1205	556	2665
	TOTAL	324	407	734	731
		100.0%	100.0%		

The results are virtually unchanged. Among those under age 50, single people are more likely than married people to go to a concert (V = 0.147**).

So far, we have looked at activities outside the home and have found that single people are somewhat more likely to eat out, go to a movie, or go to a concert. With all of these activities going on, do single people spend less time on activities around the home, such as watching television or surfing the Internet?

Data File: **GSS**
Task: **Cross-tabulation**
➤ Row Variable: **82) WATCH TV**
➤ Column Variable: **7) SINGLE/MAR**
➤ View: **Tables**
➤ Display: **Column %**

WATCH TV   by   SINGLE/MAR
Cramer's V: 0.053 *

| | | SINGLE/MAR | | | |
		SINGLE	MARRIED	Missing	TOTAL
WATCH TV	0-1 HOURS	283	549	273	832
		28.0%	28.3%		28.2%
	2-3 HOURS	432	913	516	1345
		42.7%	47.0%		45.5%
	4+ HOURS	297	480	423	777
		29.3%	24.7%		26.3%
	Missing	363	682	438	1483
	TOTAL	1012	1942	1650	2954
		100.0%	100.0%		

If you are continuing from the previous table, don't forget to delete the subset variable.

It appears that those who are single and those who are married spend about the same amount of time watching television. The percentage of single (28 percent) and married (28.3 percent) who watch less than 1 hour a day is virtually identical. If we look at the bottom row, we do see that 29.3 percent of singles watch television 4 or more hours per day compared to 24.7 percent of those who are married. So, singles are somewhat more likely to watch a lot of television, but the difference is not very great.

Data File: **GSS**
Task: **Cross-tabulation**
➤ Row Variable: **83) WEB HOURS**
➤ Column Variable: **7) SINGLE/MAR**
➤ View: **Tables**
➤ Display: **Column %**

WEB HOURS   by   SINGLE/MAR
Cramer's V: 0.103 *

| | | SINGLE/MAR | | | |
		SINGLE	MARRIED	Missing	TOTAL
WEB HOURS	1-2 HOURS	77	161	68	238
		33.3%	40.8%		38.0%
	3-5 HOURS	59	111	47	170
		25.5%	28.1%		27.2%
	6 AND UP	95	123	55	218
		41.1%	31.1%		34.8%
	Missing	1144	2229	1480	4853
	TOTAL	231	395	1650	626
		100.0%	100.0%		

There is more of a pattern in this table, which compares hours spent on the Internet. The variable 83) WEB HOURS asks respondents who use the Internet how many hours per week they typically spend surfing the Web. What we see is that single respondents (41.1 percent) who use the Internet are more likely than married respondents (31.1 percent) to spend 6 or more hours per week on the Web. So, although single respondents are more likely to be engaged in social activities outside the home, they also are somewhat more likely to spend time watching television and surfing the Internet at home.

We have seen that the singles lifestyle is a mix of activities both inside and outside the home. Does this mean that those who are single find life to be more exciting than those who are married? Let's test the hypothesis that **those who have never married will be more likely than those who are married to believe life is exciting.**

Data File: **GSS**

Task: **Cross-tabulation**

➤ Row Variable: **77) LIFE**

➤ Column Variable: **7) SINGLE/MAR**

➤ View: **Tables**

➤ Display: **Column %**

LIFE   by   SINGLE/MAR
Cramer's V: 0.093 **

		SINGLE/MAR			
		SINGLE	MARRIED	Missing	TOTAL
L I F E	EXCITING	417	837	437	1254
		46.5%	49.2%		48.3%
	ROUTINE	423	820	571	1243
		47.2%	48.2%		47.9%
	DULL	56	43	94	99
		6.3%	2.5%		3.8%
	Missing	479	924	548	1951
	TOTAL	896	1700	1650	2596
		100.0%	100.0%		

The results here are not consistent with the hypothesis. In fact, those who are married (49.2 percent) are slightly more likely than those who are single (46.5 percent) to indicate that their life is exciting. However, such a small difference is not very substantive. Overall, we would have to say that married and single adults are about equally as likely to view their life as exciting. Despite their greater involvement in outside activities, singles are not more likely than those who are married to describe their life as exciting.

Of course, lifestyles consist of more than just eating out or going to a movie or a concert—they also are comprised of our personal relationships with others. Let's see how single adults compare to those who are married when it comes to spending time with others. We'll begin with time spent with extended family.

Data File: **GSS**

Task: **Cross-tabulation**

➤ Row Variable: **78) SOC.KIN**

➤ Column Variable: **7) SINGLE/MAR**

➤ View: **Tables**

➤ Display: **Column %**

SOC.KIN   by   SINGLE/MAR
Cramer's V: 0.096 **

		SINGLE/MAR			
		SINGLE	MARRIED	Missing	TOTAL
S O C . K I N	DAILY/WKLY	494	963	555	1457
		54.3%	55.2%		54.9%
	MONTH/YEAR	357	738	453	1095
		39.3%	42.3%		41.3%
	NEVER	58	44	66	102
		6.4%	2.5%		3.8%
	Missing	466	879	576	1921
	TOTAL	909	1745	1650	2654
		100.0%	100.0%		

Based on these results, those who are single and those who are married are just as likely to visit at least weekly with members of their extended family.

Next, let's look at social time spent with neighbors.

Data File: **GSS**

Task: **Cross-tabulation**

➤ Row Variable: **79) SOC.NEIGH.**

➤ Column Variable: **7) SINGLE/MAR**

➤ View: **Tables**

➤ Display: **Column %**

SOC.NEIGH. by SINGLE/MAR
Cramer's V: 0.188 **

		SINGLE/MAR			
		SINGLE	MARRIED	Missing	TOTAL
SOC.NEIGH.	DAILY/WKLY	385	467	357	852
		42.6%	26.7%		32.1%
	MONTH/YEAR	263	819	336	1082
		29.1%	46.9%		40.8%
	NEVER	256	461	379	717
		28.3%	26.4%		27.0%
	Missing	471	877	578	1926
	TOTAL	904	1747	1650	2651
		100.0%	100.0%		

There is a difference here. Single respondents (42.6 percent) are much more likely than those who are married (26.7 percent) to socialize with neighbors at least once a week (V = .188**).

How about the frequency of socializing with friends?

Data File: **GSS**

Task: **Cross-tabulation**

➤ Row Variable: **80) SOC.FRIEND**

➤ Column Variable: **7) SINGLE/MAR**

➤ View: **Tables**

➤ Display: **Column %**

SOC.FRIEND by SINGLE/MAR
Cramer's V: 0.272 **

		SINGLE/MAR			
		SINGLE	MARRIED	Missing	TOTAL
SOC.FRIEND	DAILY/WKLY	572	610	433	1182
		63.1%	35.1%		44.7%
	MONTH/YEAR	280	1009	475	1289
		30.9%	58.0%		48.7%
	NEVER	55	121	163	176
		6.1%	7.0%		6.6%
	Missing	468	884	579	1931
	TOTAL	907	1740	1650	2647
		100.0%	100.0%		

Single people are even more likely to visit with friends. Among those who are single, 63.1 percent socialize with friends at least once a week, compared to 35.1 percent for those who are married (V = .272**).

Finally, let's look at how often those who are single and those who are married go to a bar or tavern to socialize.

Data File: **GSS**

Task: **Cross-tabulation**

➤ Row Variable: **81) SOC. BAR**

➤ Column Variable: **7) SINGLE/MAR**

➤ View: **Tables**

➤ Display: **Column %**

SOC. BAR by SINGLE/MAR
Cramer's V: 0.294 **

		SINGLE/MAR			
		SINGLE	MARRIED	Missing	TOTAL
SOC. BAR	DAILY/WKLY	278	151	148	429
		30.8%	8.7%		16.2%
	MONTH/YEAR	317	661	328	978
		35.1%	37.9%		37.0%
	NEVER	308	930	600	1238
		34.1%	53.4%		46.8%
	Missing	472	882	574	1928
	TOTAL	903	1742	1650	2645
		100.0%	100.0%		

Those who have never married (30.8 percent) are more likely than those who are married (8.7 percent) to go to a bar at least weekly. But, overall, the percentage of singles who go to a bar to socialize is much less than the percentage who visit with family, neighbors, or friends.

Being single does not mean that a person is socially isolated. On the contrary, singles are more likely than those who are married to socialize regularly with neighbors and friends or at a bar, and they are equally as likely as those who are married to socialize with extended family.

We have seen that singles are more active in social activities outside the home and that they socialize more frequently with their friends and neighbors, all of which may contribute to feelings of personal happiness. So, let's look at how those who have never married compare to those who are married with regard to overall happiness.

Data File: **GSS**
Task: **Cross-tabulation**
➤ Row Variable: **73) HAPPY?**
➤ Column Variable: **7) SINGLE/MAR**
➤ View: **Tables**
➤ Display: **Column %**

HAPPY? by SINGLE/MAR
Cramer's V: 0.259 **

		SINGLE/MAR			
		SINGLE	MARRIED	Missing	TOTAL
HAPPY?	VERY HAPPY	270	1137	365	1407
		20.0%	43.6%		35.6%
	PRET.HAPPY	876	1334	968	2210
		65.0%	51.1%		55.9%
	NOT TOO	201	138	294	339
		14.9%	5.3%		8.6%
	Missing	28	15	23	66
	TOTAL	1347	2609	1650	3956
		100.0%	100.0%		

The GSS asked respondents, "Taken all together, how would you say things are these days—would you say that you are very happy, pretty happy, or not too happy?" Looking at the results we see that those who have never married (20 percent) are less than half as likely as those who are married (43.6 percent) to describe themselves as "very happy." Those who are single (14.9 percent) are also more likely than those who are married (5.3 percent) to indicate that they are "not too happy." The difference is statistically significant (V = .259**).

Although most singles are "very happy" or "pretty happy," as a whole, levels of personal happiness are lower among singles than among those who are married. In Chapter 8, we will explore some of the other ways that marriage may be related to feelings of personal well-being. But, first, you will have the opportunity to explore another facet of the single life that we have not yet touched on—cohabitation.

Marriage and Family

WORKSHEET

CHAPTER 6

NAME: _____

COURSE: _____

DATE: _____

Workbook exercises and software are copyrighted. Copying is prohibited by law.

REVIEW QUESTIONS

Based on the first part of this chapter, answer True or False to the following items:

About one-fourth of the respondents in the GSS have never been married.	T F
Most of those who have never married indicate that they do not plan to ever get married.	T F
Historically, higher college graduation rates are associated with areas that have fewer single adults.	T F
Rural areas tend to have a higher percentage of singles.	T F
Singles are more likely than those who are married to have gone to a movie in the previous week.	T F
Those who are single are more likely than those who are married to believe that their life is exciting.	T F
Married respondents are more likely to socialize regularly with friends, neighbors, and extended family.	T F
Married respondents report higher levels of personal happiness.	T F

EXPLORIT QUESTIONS

1. The GSS asked respondents whether they agreed or disagreed with the statement that it is alright for a couple to live together without intending to get married. Take a look at how people responded.

> ➤ Data File: **GSS**
> ➤ Task: **Univariate**
> ➤ Primary Variable: **104) COHABITATE**
> ➤ View: **Pie**

a. What percentage agree that it is alright for couples to live together? _____%

b. What percentage disagree with couples living together? _____%

2. The GSS also asked respondents whether they agreed or disagreed with the statement that it is a good idea for a couple who intend to get married to live together first. Take a look at how people responded to this item.

> Data File: **GSS**
> Task: **Univariate**
> ➤ Primary Variable: **105) COHAB.FRST**
> ➤ View: **Pie**

a. What percentage agree that it is a good idea for couples to live together before getting married?   _____%

b. What percentage disagree that it is a good idea for couples to live together before getting married?   _____%

c. Which of the following statements is supported by the results in the previous two graphs? (circle one)

    1. People are more likely to agree with a couple living together if the couple is planning to get married than if they are not planning to get married.

    2. Whether or not a couple is going to get married has very little impact on whether people agree or disagree with their living together.

3. What do those who are already married think about living together first?

> Data File: **GSS**
> ➤ Task: **Cross-tabulation**
> ➤ Row Variable: **105) COHAB.FRST**
> ➤ Column Variable: **7) SINGLE/MAR**
> ➤ View: **Tables**
> ➤ Display: **Column %**

a. Fill in the percentaged results for the *top* row of the table.

	SINGLE	MARRIED
AGREE	_____%	_____%

b. What is the value of V?   V = _____

c. Is V statistically significant?   Yes   No

d. Those who are married are more likely to think a couple should live together before getting married.   T   F

4.   Let's use the auto-analyzer to look for variations in attitudes toward cohabitation.

> ➤ *Data File:* **GSS**
>   ➤ *Task:* **Auto-Analyzer**
> ➤ *Variable:* **104) COHABITATE**
>   ➤ *View:* **Age**

   a.   What percentage of those under age 30 agree that it is alright to live together?   _____%

   b.   What percentage of those age 50 or older agree that it is alright to live together?   _____%

5.   Continuing on with the auto-analyzer, select the appropriate independent variable to compare the attitudes of men and women with regard to a couple living together even if they are not going to get married.

   a.   Which of the following statements is supported by the results? (circle one)
       1.   Men are more likely to agree that it is alright for a couple to live together.
       2.   Women are more likely to agree that it is alright for a couple to live together.
       3.   Men and women are equally likely to agree that it is alright for a couple to live together.

   b.   How would you explain these results?

6.   Continuing on with the auto-analyzer, select the appropriate independent variable to compare the attitudes of blacks and whites with regard to approving of a couple living together even if they are not going to get married.

   Which of the following statements is supported by the results? (circle one)
       a.   Whites are more likely than blacks to agree that it is alright for a couple to live together.
       b.   Blacks are more likely than whites to agree that it is alright for a couple to live together.
       c.   Whites and blacks are equally likely to agree that it is alright for a couple to live together.

7.   Continuing on with the auto-analyzer, select the appropriate independent variable to compare different regions of the country with regard to a couple living together even if they are not going to get married.

   a. In which region of the country are people most likely to agree that cohabitation is alright?   _____

   b. In which region of the country are people least likely to agree that cohabitation is alright?   _____

8. Continuing on with the auto-analyzer, select the appropriate independent variable to compare different religions with regard to a couple living together even if they are not going to get married.

   a. In which religion are people least likely to agree that cohabitation is alright?   _____

   b. The attitudes of Catholics toward cohabitation closely resemble the attitudes of (circle one)
       1.   liberal Protestants.
       2.   conservative Protestants.
       3.   those who are Jewish.

9. What percentage of never-married respondents *who are in steady relationships* actually are living together?

> Data File: **GSS**
> ➤ Task: **Univariate**
> ➤ Primary Variable: **22) LIVE WITH**
> ➤ Subset Variable: **3) EVER MAR?**
> ➤ Subset Category: **Include: 2) NO**
> ➤ View: **Pie**

What percentage of respondents are living with their steady romantic partner?   _____%

10. Is there a relationship between race and cohabitation?

> Data File: **GSS**
> ➤ Task: **Cross-tabulation**
> ➤ Row Variable: **22) LIVE WITH**
> ➤ Column Variable: **26) RACE**
> ➤ Subset Variable: **3) EVER MAR?**
> ➤ Subset Category: **Include: 2) NO**
> ➤ View: **Tables**
> ➤ Display: **Column %**

   a. Fill in the percentaged results for the *top* row of the table.

	WHITE	BLACK
NO	_____%	_____%

b.   What is the value of V?   V = _____

c.   Is V statistically significant?   Yes   No

d.   Statistically speaking, blacks and whites are equally likely to be living with their
     steady romantic partner.   T   F

CHAPTER 7

MATE SELECTION

Tasks: Univariate, Cross-tabulation, Historical Trends, Auto-Analyzer
Data Files: CULTURES, GSS, TRENDS

The process of mate selection in our society seems very natural. A man and a woman are attracted to each other, they date for a period of time, they fall in love, and then they marry. Sure, there may be many variants to the story. For example, the dating process may last one week or ten years. The marriage may be rooted in convenience, security, or necessity, rather than in love. Maybe an unexpected pregnancy leads to a hasty marriage. Perhaps a couple decides to cohabit rather than go through the official process of getting married.

Anthropologists use the term "marriage" to refer to a sexual and economic relationship between individuals that is relatively permanent and can produce legitimate children. Throughout the world, more than 90 percent of people who reach adulthood marry sometime in their lifetime. But the process of mate selection varies significantly across the many cultures of the world. Begin by examining data based on preindustrial societies.

➤ *Data File:* **CULTURES**
➤ *Task:* **Univariate**
➤ *Primary Variable:* **14) MEN PICK**
➤ *View:* **Pie**

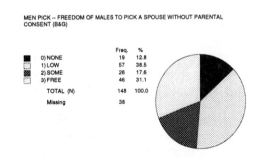

MEN PICK -- FREEDOM OF MALES TO PICK A SPOUSE WITHOUT PARENTAL CONSENT (B&G)

		Freq.	%
■	0) NONE	19	12.8
	1) LOW	57	38.5
	2) SOME	26	17.6
	3) FREE	46	31.1
	TOTAL (N)	148	100.0
	Missing	38	

In only 31.1 percent of societies are men free to pick a spouse without parental consent. In more than half of these preindustrial societies, men have little or no freedom in picking their own spouse. What about women?

113

Data File: **CULTURES**

Task: **Univariate**

➤ Primary Variable: **16) FEM.PICK**

➤ View: **Pie**

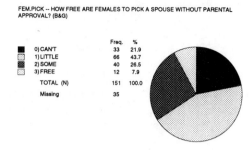

FEM.PICK -- HOW FREE ARE FEMALES TO PICK A SPOUSE WITHOUT PARENTAL APPROVAL? (B&G)

		Freq.	%
■	0) CAN'T	33	21.9
▨	1) LITTLE	66	43.7
▨	2) SOME	40	26.5
▨	3) FREE	12	7.9
	TOTAL (N)	151	100.0
	Missing	35	

Mate selection is even more restrictive for females. Only 7.9 percent of preindustrial societies provide women the freedom to pick a spouse without parental approval. In almost 22 percent of preindustrial societies, females have virtually no say in the selection of their spouse.

Must both the bride and the groom consent to marriage?

Data File: **CULTURES**

Task: **Univariate**

➤ Primary Variable: **12) COUPLE PIC**

➤ View: **Pie**

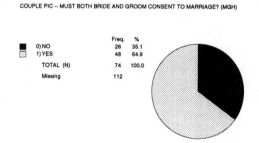

COUPLE PIC -- MUST BOTH BRIDE AND GROOM CONSENT TO MARRIAGE? (MGH)

		Freq.	%
■	0) NO	26	35.1
▨	1) YES	48	64.9
	TOTAL (N)	74	100.0
	Missing	112	

Things look a little brighter here. Although the earlier data indicate that the majority of preindustrial societies do not allow either males or females to pick a spouse without parental approval, we see that the bride and the groom must at least consent to the marriage in nearly 65 percent of societies.

The exchange of money between the families of the bride and groom was quite common in preindustrial societies—and it is still common in many societies today. The most common form of exchange is known as a **bride price**, in which a gift is given by the groom to the family of the bride. **Dowries**, which are less common, usually involve a substantial transfer of goods and money from the bride's family to the bride to bring to her husband. Several other forms of economic transactions or services relating to marriage also exist. Let's look at the percentage of societies that have some type of economic marital exchange.

Data File: **CULTURES**
Task: **Univariate**
➤ *Primary Variable:* **11) MARRIAGE $**
➤ *View:* **Pie**

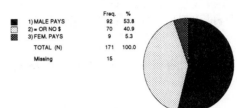

		Freq.	%
■	1) MALE PAYS	92	53.8
▦	2) = OR NO $	70	40.9
▨	3) FEM. PAYS	9	5.3
	TOTAL (N)	171	100.0
	Missing	15	

In over half of the preindustrial societies, the male pays something to the family of the bride. Payments or transfers of goods from the bride's family actually occur in very few societies (5.3 percent). These economic transactions help explain why the bride and the groom in most preindustrial societies had little control over their mate selection.

For the remainder of this chapter we will look at mate selection in the modern-day United States. Since marriages in the United States are not "arranged" by the parents of the bride and groom (although a fair amount of meddling does often occur), we will focus on how brides and grooms become attracted to one another.

Common wisdom offers us conflicting views on marital attraction. We are told there are lots of fish in the sea; at the same time, we are encouraged to find our *one* true love. We are told that "opposites attract" but also that "birds of a feather flock together." So what should we believe? Do we select our marriage partners from a wide pool of eligibles—or is the selection rather limited? Do opposites attract—or are people more attracted by similarities? One way we can address these questions is by comparing the social and demographic characteristics of marriage partners. Social scientists refer to marriage between two people with similar social characteristics as **homogamy**; marriage between two people with dissimilar backgrounds is called **heterogamy**. Let's see which of these two concepts better characterizes mate selection and marriage in our society with regard to education and religion.

If homogamy is the guiding principle behind mate selection, then a person's level of education should be an accurate predictor of his or her spouse's level of education. The hypothesis is: **People will tend to choose spouses who have the same level of education as themselves**.

➤ *Data File:* **GSS**
➤ *Task:* **Cross-tabulation**
➤ *Row Variable:* **44) MATE EDUC.**
➤ *Column Variable:* **27) EDUCATION**
➤ *Subset Variable:* **2) MARRIED?**
➤ *Subset Category:* **Include: 1) Yes**
➤ *View:* **Tables**
➤ *Display:* **Column %**

MATE EDUC.   by   EDUCATION

Cramer's V: 0.410 **

		EDUCATION				
		NO HS GRAD	HS GRAD	COLL EDUC	Missing	TOTAL
M A T E E D U C	NO HS GRAD	176	123	57	1	356
		50.4%	15.7%	4.0%		13.9%
	HS GRAD	133	406	323	2	862
		38.1%	52.0%	22.7%		33.8%
	COLL EDUC	40	252	1043	3	1335
		11.5%	32.3%	73.3%		52.3%
	Missing	17	16	26	6	65
	TOTAL	349	781	1423	12	2553
		100.0%	100.0%	100.0%		

Looking at the first column in this table we see that 50.4 percent of those who have less than a high school education married someone who also had less than a high school education. In the second row we see that 52 percent of those whose highest degree was a high school diploma married someone with a similar level of education. Skipping down to the last category, we see that 73.3 percent of those with a college education married others who had gone to college. It is apparent that people do tend to marry someone who has the same level of education as themselves (V = .410**).

Are first marriages any more or less likely than remarriages to be characterized by educational homogamy? Let's use a control variable to answer that question.

Data File: **GSS**
Task: **Cross-tabulation**
Row Variable: **44) MATE EDUC.**
Column Variable: **27) EDUCATION**
➤ Control Variable: **9) REMARRIAGE**
➤ View: **Tables: MARRIAGE**
➤ Display: **Column %**

MATE EDUC.   by   EDUCATION
Controls   REMARRIAGE: 1 MARRIAGE
Cramer's V 0.431 **

		EDUCATION				
		NO HS GRAD	HS GRAD	COLL EDUC	Missing	TOTAL
MATE EDUC.	NO HS GRAD	138	95	42	0	275
		56.1%	16.2%	3.8%		14.2%
	HS GRAD	80	305	241	1	626
		32.5%	52.0%	21.9%		32.4%
	COLL EDUC	28	186	816	2	1030
		11.4%	31.7%	74.2%		53.3%
	Missing	13	11	22	5	51
	TOTAL	246	586	1099	8	1931
		100.0%	100.0%	100.0%		

The option for selecting a control variable is found on the same screen you use to select other variables. For this example, select 9) REMARRIAGE as a control variable and then click [OK] to continue as usual. Separate tables for each of the 9) REMARRIAGE categories will now be shown for the 44) MATE EDUC and 27) EDUCATION cross-tabulation.

This first table represents the results for those who have been married only once. The pattern is the same as we saw in the previous table with the highest percentage in each column corresponding to the same educational level in the row variable. The value of Cramer's V is a very strong .431**.

Data File: **GSS**
Task: **Cross-tabulation**
Row Variable: **44) MATE EDUC.**
Column Variable: **27) EDUCATION**
Control Variable: **9) REMARRIAGE**
➤ View: **Tables: REMARRIAGE**
➤ Display: **Column %**

MATE EDUC.   by   EDUCATION
Controls: REMARRIAGE: REMARRIAGE
Cramer's V 0.356 **

		EDUCATION				
		NO HS GRAD	HS GRAD	COLL EDUC	Missing	TOTAL
MATE EDUC.	NO HS GRAD	36	27	15	1	78
		36.4%	14.4%	4.8%		13.0%
	HS GRAD	52	99	81	1	232
		52.5%	52.7%	25.9%		38.7%
	COLL EDUC	11	62	217	1	290
		11.1%	33.0%	69.3%		48.3%
	Missing	4	5	2	1	12
	TOTAL	99	188	313	4	600
		100.0%	100.0%	100.0%		

Click the appropriate button at the bottom of the task bar to look at the second (or next) partial table for 9) REMARRIAGE.

Educational homogamy has a similar level on remarriages as on first marriages. The value of Cramer's V is .356**. Whether it's a first marriage or a remarriage, people tend to marry someone who has the same level of education as themselves.

Do the parents of the GSS respondents also have similar educational backgrounds? Let's go back a generation and see if these findings still hold true.

Marriage and Family

<table>
<tr><td rowspan="2" colspan="2"></td><td align="right">Data File:</td><td>GSS</td></tr>
</table>

Data File: **GSS**
Task: **Cross-tabulation**
➤ Row Variable: **42) DAD EDUC.**
➤ Column Variable: **43) MOM EDUC.**
➤ View: **Tables**
➤ Display: **Column %**

DAD EDUC.   by   MOM EDUC.
Cramer's V: 0.383 **

		MOM EDUC.				
		NO HS GRAD	HS GRAD	COLL EDUC	Missing	TOTAL
D A D E D U C	NO HS GRAD	895	363	183	87	1441
		62.4%	20.8%	12.2%		30.8%
	HS GRAD	193	722	261	58	1176
		13.5%	41.3%	17.3%		25.1%
	COLL EDUC	346	662	1061	177	2069
		24.1%	37.9%	70.5%		44.2%
	Missing	154	124	59	304	641
	TOTAL	1434	1747	1505	626	4686
		100.0%	100.0%	100.0%		

Remember to remove the control variable before continuing.

The trend here also is consistent with the principle of homogamy. In each of the columns the highest percentage is for someone with the same level of education. Among those with less than a high school education, 62.4 percent married people who also had not completed high school. For those with a high school diploma, 41.3 percent married people with the same level of education. Among those who went to college, 70.5 percent have an educationally homogamous marriage. The value of Cramer's V is almost as strong in this table as it was in the previous table on the respondents' own marriages (V = .383**). The influence of homogamy is as evident in the previous generation as it is in the current generation.

Many factors contribute to the high incidence of educational homogamy. One would be opportunity. Many people meet their future spouses while attending school—either in high school or beyond. The influence of age homogamy would be another factor. People of a similar age are more likely to have the same level of education. Yet another factor would be childhood socialization. People who have similar levels of education probably come from homes with similar socioeconomic characteristics, where similar educational values are taught. Then there is the socialization that takes place in college where opinions and attitudes are shaped even more. Finally, on the interpersonal level, people tend to feel more comfortable with those who they feel are similar to themselves. Such interactions are more predictable and also reinforcing of our own world view.

Next, let's see if the same trend toward homogamy is evident with regard to religious affiliations. Did the respondent's mother and father tend to share the same religious identity?

Data File: **GSS**
Task: **Cross-tabulation**
➤ Row Variable: **53) POPS RELIG**
➤ Column Variable: **52) MOMS RELIG**
➤ View: **Tables**
➤ Display: **Column %**

POPS RELIG   by   MOMS RELIG
Cramer's V: 0.853 **

		MOMS RELIG				
		PROTESTANT	CATHOLIC	JEWISH	Missing	TOTAL
P O P S R E L I G	PROTESTANT	552	42	0	7	594
		93.1%	12.3%	0.0%		62.2%
	CATHOLIC	38	298	0	8	336
		6.4%	87.1%	0.0%		35.2%
	JEWISH	3	2	20	0	25
		0.5%	0.6%	100.0%		2.6%
	Missing	111	50	2	4516	4679
	TOTAL	593	342	20	4531	955
		100.0%	100.0%	100.0%		

The results are consistent with the principle of homogamy. By far the largest percentage in each column is for those who are married to people with similar religious orientations—93.1 percent for

Protestants, 87.1 percent for Catholics, 100 percent for those who are Jewish. The marriages of most of the respondents' parents are characterized by religious homogamy (V = .853**). However, we must note that we don't know whether this homogamy occurred before the marriage or after. It is possible that one of the parents changed their religious identity to match that of their spouse after they got married. But it seems likely that religious homogamy probably played a role in the initial attraction as well.

Of course, not everyone who identifies with a particular religion attaches the same level of importance to their beliefs. By looking at actual church attendance rather than just religious affiliation, we can get a sense of the extent to which a respondent's parents had similar levels of religiosity.

Data File: **GSS**
Task: **Cross-tabulation**
➤ Row Variable: **56) PA ATTEND**
➤ Column Variable: **55) MA ATTEND**
➤ View: **Tables**
➤ Display: **Column %**

PA ATTEND   by   MA ATTEND
Cramer's V: 0.514 **

| | | MA ATTEND | | | | |
		RARELY	INFREQUENT	OFTEN	Missing	TOTAL
PA ATTEND	RARELY	156	90	99	9	345
		84.8%	38.6%	16.5%		34.0%
	INFREQUENT	11	120	79	7	210
		6.0%	51.5%	13.2%		20.7%
	OFTEN	17	23	421	13	461
		9.2%	9.9%	70.3%		45.4%
	Missing	12	28	59	4505	4604
	TOTAL	184	233	599	4534	1016
		100.0%	100.0%	100.0%		

This table compares the actual church attendance practices of each of the respondent's parents. Again, the influence of homogamy is evidenced by the strength of the relationship between these two variables (V = .514**). Among those whose mothers rarely attended church, 84.8 percent also had fathers who rarely attended church. And 70.3 percent of those whose mothers attended church often also had fathers who attended often. So, not only did most of the respondents have parents who shared the same religious identity, the parents also tended to have similar levels of church attendance.

Based on the analyses presented thus far, we can conclude that with regard to education and religion, most marriages are homogamous. Although we are free to marry anyone we please, most people are attracted to those who are similar to themselves. There are many fish in the sea, but most of us choose to stay in our own ponds.

In the worksheet section that follows, you will have the opportunity to explore attitudes toward another facet of this issue, and that is interracial marriages. Do Americans think there should be laws against interracial marriages? How do people think their family would react if they were to marry someone of another race? These are just a few of the questions you will have the opportunity to answer on your own.

Marriage and Family

NAME: _____

COURSE: _____

DATE: _____

REVIEW QUESTIONS

Based on the first part of this exercise, answer True or False to the following items:

Because males initiated the marriage proposal in most preindustrial societies, females had more freedom to choose a marriage partner without parental consent.	T F
In preindustrial societies, less than 10 percent of females were free to choose their spouse without parental consent.	T F
In over half of preindustrial societies, the groom made some type of payment to the family of the bride.	T F
Most people marry someone who has the same level of education as themselves.	T F
Those who have remarried are more likely than those who have been married only once to have the same level of education as their spouse.	T F
Most respondents' mothers and fathers shared the same level of education.	T F
Most respondents' mothers and fathers shared the same religious identity.	T F
The frequency of church attendance of a respondent's mother is not a good predictor of the frequency of attendance of a respondent's father.	T F

EXPLORIT QUESTIONS

1. Marriages between persons of different races are commonly referred to as interracial marriages. The GSS asked respondents about interracial marriages between African Americans and whites in the United States. Let's explore this variable.

    a. Answer this question before doing the analysis: What percentage of respondents do you think will believe that there should be laws against interracial marriages?   _____%

      ➤ *Data File:* **GSS**
      ➤ *Task:* **Univariate**
      ➤ *Primary Variable:* **57) INTERMAR.?**
      ➤ *View:* **Pie**

    b. What percentage of respondents believe there should be laws against interracial marriage between African Americans and whites?   _____%

Chapter 7: Mate Selection

c. Is this percentage higher or lower than you predicted?   Higher   Lower

2. The GSS has included the intermarriage question in almost every survey since 1972. How have attitudes toward interracial marriage changed over this period of time?

> ➤ *Data File:* **TRENDS**
> ➤ *Task:* **Historical Trends**
> ➤ *Variable:* **31) INTERMAR.?**

a. According to the description for this variable, this graph represents the percentage of GSS respondents who believe there should be laws against interracial marriage.   Yes   No

b. Describe the trend presented in this graph in two or three sentences.

3. For the following questions, use the auto-analyzer to look for variations in attitudes toward interracial marriage.

> ➤ *Data File:* **GSS**
> ➤ *Task:* **Auto-Analyzer**
> ➤ *Variable:* **57) INTERMAR.?**
> ➤ *View:* **Age**

a. There is not a statistically significant relationship between age and attitudes toward interracial marriage.   T   F

b. The attitudes of those age 30–49 most closely resemble (circle one)
    1. the attitudes of those under 30.
    2. the attitudes of those age 50 or older.

c. Are the results in this graph consistent with the trend in Question 2? Explain your answer.

4. How does education influence attitudes toward interracial marriage?

> *Data File:* **GSS**
> *Task:* **Auto-Analyzer**
> *Variable:* **57) INTERMAR.?**
> ➤ *View:* **Education**

   a. Fill in the percentaged results for the *top* row of the table.

	NO HS GRAD	HS GRAD	COLL EDUC
YES	_____%	_____%	_____%

   b. Those who did not graduate from college are about _____ times as likely than those who went to college to believe there should be laws against interracial marriage.

   _____

5. Do men and women share similar attitudes toward interracial marriage?

> *Data File:* **GSS**
> *Task:* **Auto-Analyzer**
> *Variable:* **57) INTERMAR.?**
> ➤ *View:* **Sex**

   There is not a statistically significant difference between men and women on this issue.   T   F

6. What is the relationship between race and attitudes toward interracial marriage?

> *Data File:* **GSS**
> *Task:* **Auto-Analyzer**
> *Variable:* **57) INTERMAR.?**
> ➤ *View:* **Race**

   a. Fill in the percentaged results for the *top* row of the table.

	WHITE	BLACK
YES	_____%	_____%

   b. There is not a statistically significant difference between blacks and whites on this issue.   T   F

7. Does the region of the country in which a respondent lives affect his or her attitude toward interracial marriage?

> *Data File:* **GSS**
> *Task:* **Auto-Analyzer**
> *Variable:* **57) INTERMAR.?**
> ➤ *View:* **Region**

   a. Which region of the country has the lowest percentage of respondents who support laws against interracial marriage?   _____

   b. Which region of the country has the highest percentage of respondents who support laws against interracial marriage?   _____

   c. Are the regional differences statistically significant?   Yes   No

8. What is the relationship between religion and attitudes toward interracial marriage?

> *Data File:* **GSS**
> *Task:* **Auto-Analyzer**
> *Variable:* **57) INTERMAR.?**
> ➤ *View:* **Religion**

   a. Fill in the percentaged results for the *top* row of the table.

	LIB. PROT.	CON. PROT.	CATHOLIC	JEWISH
YES	_____%	_____%	_____%	_____%

   a. Which religion has the lowest percentage of respondents who support laws against interracial marriage?   _____

   b. Which religion has the highest percentage of respondents who support laws against interracial marriage?   _____

   c. Are the religious differences statistically significant?   Yes   No

9. Finally, let's look at political party affiliation.

   a. Before looking at the results, what do you expect to find? (circle one)
      1. Republicans will be more likely to support laws against interracial marriage.
      2. Democrats will be more likely to support laws against interracial marriage.
      3. There will not be a statistically significant relationship between political party affiliation and attitudes toward interracial marriage.

         Data File: **GSS**
            Task: **Auto-Analyzer**
       Variable: **57) INTERMAR.?**
       ➤ *View:* **Party**

b. According to these results: (circle one)
   1. Republicans are more likely to support laws against interracial marriage.
   2. Democrats are more likely to support laws against interracial marriage.
   3. There is not a statistically significant relationship between political party affiliation and attitudes toward interracial marriage.

10. Summarize the results from Questions 1–9. Include a description of overall attitudes toward interracial marriage as well as variations between different groups for *each* of the different independent variables.

11. Attitudes about laws against interracial marriage do not really tell us how people think their family would respond if someone in their family were to marry someone from a different race. Let's see how whites say they would respond to someone in their own family marrying an African American.

              Data File: **GSS**
            ➤ *Task:* **Univariate**
   ➤ *Primary Variable:* **58) MARRY BLK**
   ➤ *Subset Variable:* **26) RACE**
   ➤ *Subset Category:* **Include: 1) WHITE**
             ➤ *View:* **Pie**

a. What percentage of white respondents would be opposed to someone in their family marrying an African American?   _____%

b. Looking back at Question 6, what percentage of white respondents believe there should be laws against interracial marriage?   _____%

c. Among whites, the percentage who would oppose an interracial marriage in their family is about _____ times as high as the percentage who believe interracial marriages should be against the law.

_____

d. How would you explain these results?

Mate selection 113
               page
5

Child rearing &
Socializat pg 179
6

CHAPTER 8

THE VALUE OF FAMILY

Tasks: Univariate, Cross-tabulation, Historical Trends, Mapping
Data Files: GSS, TRENDS, GLOBAL

How important are family ties in modern societies where starting a family or even maintaining relationships with extended family is a choice rather than a necessity? Do people still feel it is important to stay in touch with members of their extended family? What place does even marriage have in a modern culture where individuals are encouraged to focus on their own needs and desires? Those are a few of the questions we will try to answer in this chapter as we explore the value of marriage and the family in contemporary societies.

Industrialization has done more than simply change the type of work we do on a daily basis. The transition from an agrarian to an industrial society also affects where we choose to live. In agrarian societies, people tend to stay close to their roots because they are dependent on the land for their survival. Even if land is available, most children will not find it practical to move far from home to start a farm of their own with no one to share the labor. The best economic opportunities are at home, or at least in the same community where they grew up. In this setting, the *extended* family (grandparents, cousins, and so on) is a valuable source of both economic and social support. Tangible assets such as land or tools can be shared, along with intangible assets such as knowledge, companionship, or assistance with child care.

Industrialization dramatically alters the distribution of the population and thus affects the structure of the family as well. Industrialization is inevitably accompanied by urbanization. With industrialization, economic opportunities lie not at home but in the marketplace. As agriculture becomes more productive and less labor-intensive, the demand for workers shifts from the farm to the factory. Thus, young people begin looking to the city for their future.

The *nuclear* family, which consists of a married couple and any children they may have, is compatible with life in an industrialized, urbanized society. Extended family ties may be a hindrance to economic and social opportunity if they keep one from pursuing other possibilities. The reduction of responsibilities to the extended family allows children the freedom to move away when they get old enough to pursue their own interests, such as going to college or taking a job in another state. The economy benefits by having a trained and mobile workforce, while individuals benefit from increased personal freedom.

In light of the centrality of the nuclear family, let's look at how often Americans socialize with their extended family.

➤ *Data File:* **GSS**
➤ *Task:* **Univariate**
➤ *Primary Variable:* **78) SOC.KIN**
➤ *View:* **Pie**

SOC.KIN -- HOW OFTEN: Spend a social evening with relatives? (SOCREL)

	Freq.	%
1) DAILY/WKLY	2012	54.0
2) MONTH/YEAR	1548	41.5
3) NEVER	168	4.5
TOTAL (N)	3728	100.0
Missing	1921	

This graph shows how often people say they spend an evening socializing with relatives. The smallest category is those who say they never socialize with relatives (4.5 percent). On the other hand, 54 percent say they socialize with family at least once per week, and an additional 41.5 percent socialize with family at least once per year. So it is safe to say that most people still believe that it is important to spend time with the extended family.

How do these percentages compare with how often people socialize with their neighbors or their friends? Neighbors, by definition, live nearby whereas relatives may live across town or in another part of the country. So, are people more likely to socialize with neighbors than with family?

Data File: **GSS**
Task: **Univariate**
➤ *Primary Variable:* **79) SOC.NEIGH.**
➤ *View:* **Pie**

SOC.NEIGH. -- HOW OFTEN: Spend a social evening with someone who lives in your neighborhood? (SOCOMMUN)

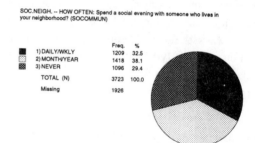

	Freq.	%
1) DAILY/WKLY	1209	32.5
2) MONTH/YEAR	1418	38.1
3) NEVER	1096	29.4
TOTAL (N)	3723	100.0
Missing	1926	

This graph shows how often people say they spend an evening socializing with neighbors. Whereas only 4.5 percent of those surveyed said they never socialize with family, 29.4 percent say they never socialize with their neighbors. Only 32.5 percent of respondents say they socialize at least weekly with their neighbors, which compares to 54 percent who say they socialized that frequently with relatives.

Of course, not everyone is friends with his or her neighbors. So, we also need to look at how frequently people spend an evening with friends who don't live in their neighborhood.

Marriage and Family

Data File: **GSS**
Task: **Univariate**
➤ Primary Variable: **80) SOC.FRIEND**
➤ View: **Pie**

SOC.FRIEND -- HOW OFTEN: Spend a social evening with friends who live outside the neighborhood? (SOCFREND)

		Freq.	%
■	1) DAILY/WKLY	1615	43.4
▨	2) MONTH/YEAR	1764	47.4
▩	3) NEVER	339	9.1
	TOTAL (N)	3718	100.0
	Missing	1931	

This graph appears to be somewhat similar to the graph depicting how frequently people socialize with relatives. Nearly half of the respondents (43.4 percent) say they spend at least one evening a week socializing with their friends. Another 47.4 percent socialize with friends monthly or annually. Only 9.1 percent of the respondents say they never spend an evening with friends. Still, this percentage is higher than the 4.5 percent of the respondents who say they never socialize with relatives. Overall, people tend to socialize with relatives more than they do with friends and neighbors. These results suggest that modern Americans still choose to value their ties with extended family.

Let's look back over a couple of decades to see if there has been any change in the percentage of people who socialize with family at least once per week.

➤ Data File: **TRENDS**
➤ Task: **Historical Trends**
➤ Variable: **19) SOC.KIN%**

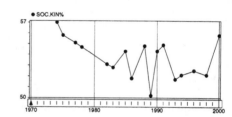

Percentage of respondents who socialize with kin at least weekly

This line graph illustrates the percentage of respondents who socialize with kin at least weekly, from 1972 to 2000. At first, it looks as though there has been considerable fluctuation in this variable over time. However, a closer look reveals that the changes that have occurred are not very dramatic. If you look at the scale on the left side of the graph, you will see that the range varies only 7 points— from 50 to 57 percent. In most years the change is less than 2 percent, and there is not a consistent decline or increase over time. There actually has not been a very noticeable change in the frequency of visits with family since 1972. (This is a good example of why it is important to always check the scale whenever you are analyzing a graph. Even small changes can be made to look big depending on the way the data are presented.)

Let's turn our attention to explaining some of the variation that does occur in the socializing with relatives variable. Why are some people more likely than others to visit with extended family? The one factor that probably inhibits socialization with kin more than any other is geographic mobility. Take a look at just how mobile we are in modern society.

> *Data File:* **GSS**
> *Task:* **Univariate**
> *Primary Variable:* **47) MOVERS**
> *View:* **Pie**

MOVERS -- When you were 16 years old, were you living in this same (city/town/county)? (MOBILE16)

		Freq.	%
■	1) SAME AREA	2201	39.2
▨	2) MOVED	3417	60.8
	TOTAL (N)	5618	100.0
	Missing	31	

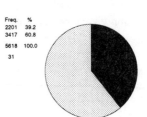

A little over one-third (39.2 percent) of the respondents live in the same city that they lived in when they were 16 years old. Most people have moved away from their childhood homes. So, how does this mobility affect how often people socialize with extended family?

> *Data File:* **GSS**
> *Task:* **Cross-tabulation**
> *Row Variable:* **78) SOC.KIN**
> *Column Variable:* **47) MOVERS**
> *View:* **Tables**
> *Display:* **Column %**

SOC.KIN by MOVERS
Cramer's V: 0.150 **

		MOVERS			
		SAME AREA	MOVED	Missing	TOTAL
SOC.KIN	DAILY/WKLY	905	1093	14	1998
		63.3%	48.0%		53.9%
	MONTH/YEAR	472	1069	7	1541
		33.0%	46.9%		41.6%
	NEVER	52	116	0	168
		3.6%	5.1%		4.5%
	Missing	772	1139	10	1921
	TOTAL	1429	2278	31	3707
		100.0%	100.0%		

Geographic mobility does affect how much we interact with extended family, but not as much as you might have expected. Reading across the top category, we see that among those who live in the same place they did as an adolescent, 63.3 percent visit with relatives at least once a week, compared to 48 percent for those who have moved to a different city. So nearly half of those who have moved to a different city still see their extended family at least once a week. Even among those who have moved, only a handful of people say they never visit with relatives (5.1 percent). For some who have moved, the less frequent interaction with extended family may be indicative of the quality of those relationships. For others, however, the less frequent interaction is just a by-product of living in a different community.

How about the influence of race on extended family relationships? Is there a difference between blacks and whites with regard to how frequently they visit with extended family?

128

Data File: **GSS**

Task: **Cross-tabulation**

Row Variable: **78) SOC.KIN**

➤ Column Variable: **26) RACE**

➤ Control Variable: **47) MOVERS**

➤ View: **Tables (Same area)**

➤ Display: **Column %**

SOC.KIN by RACE
Controls: MOVERS: SAME AREA
Cramer's V: 0.076 *

		RACE			
		WHITE	BLACK	Missing	TOTAL
SOC.KIN	DAILY/WKLY	679	180	46	859
		62.5%	65.2%		63.0%
	MONTH/YEAR	374	79	19	453
		34.4%	28.6%		33.2%
	NEVER	34	17	1	51
		3.1%	6.2%		3.7%
	Missing	599	140	33	772
	TOTAL	1087	276	99	1363
		100.0%	100.0%		

The option for selecting a control variable is found on the same screen you use to select other variables. For this example, select 47) MOVERS as a control variable and then click [OK] to continue as usual. Separate tables for each of the 47) MOVERS categories will now be shown for the 78) SOC.KIN and 26) RACE cross-tabulation.

This first table includes only those who live in the same community as when they were an adolescent. Overall, there is just a very small difference between the percentages of whites (62.5 percent) and blacks (65.2 percent) who socialize weekly with their extended family. Let's look now at those who have moved.

Data File: **GSS**

Task: **Cross-tabulation**

Row Variable: **78) SOC.KIN**

Column Variable: **26) RACE**

Control Variable: **47) MOVERS**

➤ View: **Tables (Moved)**

➤ Display: **Column %**

SOC.KIN by RACE
Controls: MOVERS: MOVED
Cramer's V: 0.059 *

		RACE			
		WHITE	BLACK	Missing	TOTAL
SOC.KIN	DAILY/WKLY	863	154	76	1017
		47.0%	53.8%		47.9%
	MONTH/YEAR	888	114	67	1002
		48.3%	39.9%		47.2%
	NEVER	87	18	11	105
		4.7%	6.3%		4.9%
	Missing	932	124	83	1139
	TOTAL	1838	286	237	2124
		100.0%	100.0%		

Click the appropriate button at the bottom of the task bar to look at the second (or next) partial table for 47) MOVERS.

Among those who have moved, the differences are slightly greater; 53.8 percent of African Americans still visit at least weekly with extended family compared to 47 percent of whites. So, the difference is very small, but blacks may be somewhat more likely than whites to maintain relationships with their extended family (V = .059*). (Remember, with a very large sample like this one, it is not unusual for even very small relationships to be statistically significant.)

If socializing weekly with relatives is indicative of the importance people place on family, then African Americans and whites are fairly similar in this regard. But how do most people perceive the commitment that different races have to families in general? Do most people think that blacks and whites share strong commitments to their families?

Data File: **GSS**
➤ Task: **Univariate**
➤ Primary Variable: **60) FAMBLKS**
➤ View: **Pie**

FAMBLKS -- BLACKS COMMITTED TO STRONG FAMILIES: Where would you rate
Blacks in general on this scale? (FAMBLKS)

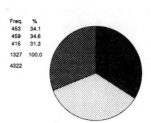

		Freq.	%
■	1) HIGH	453	34.1
▦	2) MEDIUM	459	34.6
▨	3) LOW	415	31.3
	TOTAL (N)	1327	100.0
	Missing	4322	

The GSS asked respondents how different racial and ethnic groups rank on a scale of commitment to strong families. According to this graph, the respondents were evenly divided on the answer to this question with regard to African Americans. About one-third (34.1 percent) place African Americans high on the scale of commitment to strong families, 34.6 percent rate African Americans as medium on this scale, and 31.3 percent rank them as low. How do these percentages compare with those for the same question asked about whites?

Data File: **GSS**
Task: **Univariate**
➤ Primary Variable: **59) FAMWHTS**
➤ View: **Pie**

FAMWHTS -- WHITES COMMITTED TO STRONG FAMILIES: Where would you rate
Whites in general on this scale? (FAMWHTS)

		Freq.	%
■	1) HIGH	720	53.8
▦	2) MEDIUM	434	32.4
▨	3) LOW	185	13.8
	TOTAL (N)	1339	100.0
	Missing	4310	

Overall, the respondents believe whites are more committed to strong families. Over half (53.8 percent) say that whites are high on the family commitment scale, and only 13.8 percent rank them as low. Of course, there are more white than black respondents in this sample. Let's see if the race of the respondent influenced how people answered the previous two questions.

Data File: **GSS**
➤ Task: **Cross-tabulation**
➤ Row Variable: **60) FAMBLKS**
➤ Column Variable: **26) RACE**
➤ View: **Tables**
➤ Display: **Column %**

FAMBLKS   by   RACE
Cramer's V: 0.164 **

		RACE			
		WHITE	BLACK	Missing	TOTAL
FAMBLKS	HIGH	340	98	15	438
		32.0%	49.7%		34.7%
	MEDIUM	364	69	26	433
		34.2%	35.0%		34.3%
	LOW	360	30	25	390
		33.8%	15.2%		30.9%
	Missing	3415	635	272	4322
	TOTAL	1064	197	338	1261
		100.0%	100.0%		

The race of the respondent does appear to have influenced how people answered the question pertaining to the commitment of African Americans to strong families. Blacks (49.7 percent) are much more likely than whites (32 percent) to believe that African Americans are highly committed to strong families. And blacks (15.2 percent) are half as likely as whites (33.8 percent) to believe that African Americans score low on this scale. The difference is statistically significant (V = .164**).

Let's look at the perceptions people have about other racial and ethnic groups.

Data File: **GSS**
➤ Task: **Univariate**
➤ Primary Variable: **62) FAMHSPS**
➤ View: **Pie**

FAMHSPS -- HISPANICS COMMITTED TO STRONG FAMILIES: Where would you rate Hispanics in general on this scale? (FAMHSPS)

		Freq.	%
■	1) HIGH	719	57.1
▨	2) MEDIUM	369	29.3
▨	3) LOW	171	13.6
	TOTAL (N)	1259	100.0
	Missing	4390	

The results for Hispanics are similar to the results for whites. Among those surveyed, 57.1 percent rate Hispanics high on the family commitment scale, 29.3 percent medium, and 13.6 percent low.

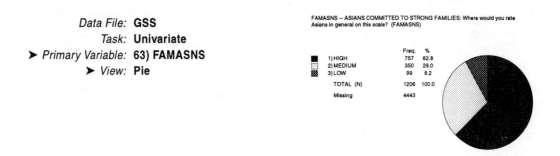

Data File: **GSS**
Task: **Univariate**
➤ Primary Variable: **63) FAMASNS**
➤ View: **Pie**

FAMASNS -- ASIANS COMMITTED TO STRONG FAMILIES: Where would you rate Asians in general on this scale? (FAMASNS)

		Freq.	%
■	1) HIGH	757	62.8
▨	2) MEDIUM	350	29.0
▨	3) LOW	99	8.2
	TOTAL (N)	1206	100.0
	Missing	4443	

Asians are perceived to be even a little more committed to strong families. Almost two-thirds (62.8 percent) of respondents rated Asians high on the family commitment scale and only 8.2 percent rated them as low.

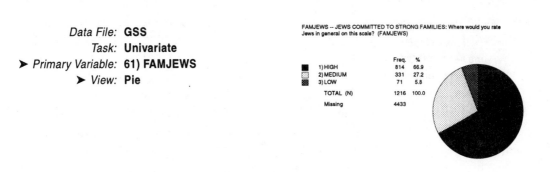

Data File: **GSS**
Task: **Univariate**
➤ Primary Variable: **61) FAMJEWS**
➤ View: **Pie**

FAMJEWS -- JEWS COMMITTED TO STRONG FAMILIES: Where would you rate Jews in general on this scale? (FAMJEWS)

		Freq.	%
■	1) HIGH	814	66.9
▨	2) MEDIUM	331	27.2
▨	3) LOW	71	5.8
	TOTAL (N)	1216	100.0
	Missing	4433	

The belief that Jews are committed to strong families is even greater than that for Asians. Among those surveyed, 66.9 percent rate Jews as high on the family commitment scale, 27.2 percent medium, and 5.8 percent low.

Although the above variables do not measure actual commitment to strong families, they do give us a sense of how people perceive family commitment among these groups. Such stereotypes may influence the way people interact with members of these different racial and ethnic groups. Because the family is generally held in such high regard in our culture, those who are stereotyped as being less committed to strong families are likely to be viewed as less admirable than those who are perceived otherwise. Of course, it also could be that people attribute anti-family labels to groups they are prejudiced against for other reasons.

Let's turn our focus now from the family to marriage in particular. At the beginning of this chapter, we noted how modern societies, such as the United States, tend to be more individualistic. Individuals are given a great deal of freedom to pursue their individual interests and take care of their own needs. So how does marriage fare in these changing times? In previous chapters, we have seen that there is a general decrease in the percentage of Americans who are married. Does this mean that people in the United States think marriage is "old fashioned" or just passé? Let's compare the attitudes of people in the United States with those from other countries on whether they think that marriage is an outdated institution.

> ➤ *Data File:* **GLOBAL**
>   ➤ *Task:* **Mapping**
> ➤ *Variable 1:* **48) WED PASSE'**
>   ➤ *View:* **Map**

WED PASSE' -- PERCENT WHO AGREE THAT "MARRIAGE IS AN OUTDATED INSTITUTION." (WVS)

Do these results surprise you? The United States has a very light color, which suggests that a low percentage of people—compared to other nations—think that marriage is an outdated institution. Let's look at the actual ranking.

    Data File: **GLOBAL**
      Task: **Mapping**
 Variable 1: **48) WED PASSE'**
    ➤ *View:* **List: Rank**

WED PASSE': Percent who agree that "marriage is an outdated institution."

RANK	CASE NAME	VALUE
1	France	29
2	Brazil	27
3	Belgium	23
4	Portugal	22
5	Netherlands	21
6	United Kingdom	18
6	Denmark	18
6	Slovenia	18
9	Mexico	17
10	Belarus	16

Marriage and Family

France leads the list with 29 percent of people believing that marriage is an outdated institution. It is followed by Brazil (27 percent), Belgium (23 percent), and Portugal (22 percent). At the bottom of the list we find India (5 percent), Iceland (6 percent), Japan (7 percent), and the United States (8 percent). When compared to the rest of the world, people in the United States are fairly traditional when it comes to the institution of marriage.

So, most people do not view marriage as an outdated institution, but what difference does being married actually make in people's lives? In the worksheet pages that follow, you will have the opportunity to address this question by exploring how marriage is related to income, health, happiness, and other measures of a person's lifestyle and attitudes.

WORKSHEET

CHAPTER

8

NAME: _____

COURSE: _____

DATE: _____

Workbook exercises and software are copyrighted. Copying is prohibited by law.

REVIEW QUESTIONS

Based on the first part of this chapter, answer True or False to the following items:

More than half of all Americans socialize with their extended family at least weekly.	T	F
Since 1972, there has been a dramatic decrease in the frequency of visits with extended family.	T	F
Moving out of state does not have a significant effect on how often people socialize with their extended families.	T	F
Whites and Hispanics are perceived to be the most committed to strong families.	T	F
Whites are more likely than blacks to rate African Americans low on a scale of commitment to strong families.	T	F
Compared to people in other countries, more Americans believe that marriage is an outdated institution.	T	F

EXPLORIT QUESTIONS

1. The 2) MARRIED? variable allows us to compare those who are married with those who, for whatever reason, are not married. Let's use this variable to see how marriage is related to family income.

> ➤ Data File: **GSS**
> ➤ Task: **Cross-tabulation**
> ➤ Row Variable: **28) INCOME**
> ➤ Column Variable: **2) MARRIED?**
> ➤ View: **Tables**
> ➤ Display: **Column %**

    a. Fill in the percentaged results for this table.

	YES	NO
LOW	_____%	_____%
MIDDLE	_____%	_____%
HIGH	_____%	_____%

b. What is the value of V?

V = _____

c. Is V statistically significant?

Yes   No

d. Those who are not married are _____ times as likely to be in the lowest family income category.

_____

e. Those who are married are _____ times as likely to be in the highest family income category.

_____

2. The 28) INCOME variable includes the incomes of everyone in the household. Is marriage related to a respondent's individual income as well?

> Data File: **GSS**
> Task: **Cross-tabulation**
> ➤ Row Variable: **29) R.INCOME**
> ➤ Column Variable: **2) MARRIED?**
> ➤ View: **Tables**
> ➤ Display: **Column %**

a. Fill in the percentaged results for this table.

	YES	NO
$0K–$17.4K	_____ %	_____ %
$17.5K–$34.9K	_____ %	_____ %
$35K +	_____ %	_____ %

b. What is the value of V?

V = _____

c. Is V statistically significant?

Yes   No

d. Being married is statistically associated with (circle one)
    1. higher levels of family income.
    2. higher levels of individual income.
    3. higher levels of both family and individual income.

e. With regard only to individual income, which variable do you think is the independent variable and which is the dependent variable in this relationship? Does being married tend to lead to higher personal incomes, or does having a high personal income make it more likely that someone will be married? Explain your answer.

3. Is being married related to a person's assessment of his or her physical health?

> Data File: **GSS**
> Task: **Cross-tabulation**
> ➤ Row Variable: **76) HEALTH**
> ➤ Column Variable: **2) MARRIED?**
> ➤ View: **Tables**
> ➤ Display: **Column %**

a. Fill in the percentaged results for the top row of the table.

	YES	NO
EXCELLENT	_____%	_____%

b. What is the value of V?   V = _____

c. Is V statistically significant?   Yes   No

d. Those who are married are somewhat more likely to rate their health as excellent.   T   F

e. Which do you think comes first, good physical health or being married? Explain your answer.

Chapter 8: The Value of Family

4. How about personal happiness? Are those who are married any more or less happy than those who are not married?

> Data File: **GSS**
> Task: **Cross-tabulation**
> ➤ Row Variable: **73) HAPPY?**
> ➤ Column Variable: **2) MARRIED?**
> ➤ View: **Tables**
> ➤ Display: **Column %**

a. Fill in the percentaged results for the *top* row of the table.

	YES	NO
VERY HAPPY	_____%	_____%

b. What is the value of V?

V = _____

c. Is V statistically significant?

Yes   No

d. Those who are married are twice as likely to indicate that they are very happy.

T   F

e. Which do you think comes first, feelings of personal happiness or being married? Explain your answer.

5. Next, let's examine the relationship between marital status and political ideology.

> Data File: **GSS**
> Task: **Cross-tabulation**
> ➤ Row Variable: **40) POL. VIEW**
> ➤ Column Variable: **2) MARRIED?**
> ➤ View: **Tables**
> ➤ Display: **Column %**

a. What percentage of those who are married describe themselves as conservative? _____%

b. What percentage of those who are not married describe themselves as conservative? _____%

c. What percentage of those who are married describe themselves as liberal? _____%

d. What percentage of those who are not married describe themselves as liberal? _____%

e. Which do you think comes first, conservative political views or being married? Explain your answer.

6. Finally, let's look at the relationship between marital status and church attendance.

> Data File: **GSS**
> Task: **Cross-tabulation**
> ➤ Row Variable: **50) ATTEND**
> ➤ Column Variable: **2) MARRIED?**
> ➤ View: **Tables**
> ➤ Display: **Column %**

a. Fill in the percentaged results for the *bottom* row of the table.

	YES	NO
WEEKLY	_____%	_____%

b. What is the value of V? V = _____

c. Is V statistically significant? Yes   No

d. Which do you think comes first, weekly church attendance or being married? (circle one)
    1. weekly church attendance
    2. marriage

7. Based on the results in this chapter, do you think marriage is beneficial to individuals and to society? Or do you believe that one's marital status is just a by-product of other things, such as personal income or religious and political views? Explain your answer.

CHAPTER 9

MARITAL HAPPINESS

Tasks: Univariate, Cross-tabulation, Historical Trends, Mapping, Correlation
Data Files: GSS, TRENDS, GLOBAL

Most people go into marriage expecting that their relationships with their spouses will be sources of personal happiness and fulfillment. Although people may believe that economic and social concerns are important, in modern societies, emotional fulfillment is usually the most important benchmark used to judge the quality of a marital relationship. People want more than just a stable relationship—they want to be happily married. But not every marriage can be described as happy. Even marriages that last a lifetime vary with regard to how much personal fulfillment each partner derives from the relationship. In this chapter, we will attempt to explain some of the variation in marital happiness and look for connections between marital happiness and other social characteristics.

Let's begin by looking at how people in the United States rate their marital happiness.

➤ *Data File:* **GSS**
    ➤ *Task:* **Univariate**
➤ *Primary Variable:* **74) HAP.MARR.?**
    ➤ *View:* **Pie**

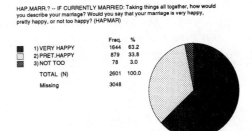

HAP.MARR.? -- IF CURRENTLY MARRIED: Taking things all together, how would you describe your marriage? Would you say that your marriage is very happy, pretty happy, or not too happy? (HAPMAR)

		Freq.	%
■	1) VERY HAPPY	1644	63.2
▨	2) PRET.HAPPY	879	33.8
▦	3) NOT TOO	78	3.0
	TOTAL (N)	2601	100.0
	Missing	3048	

The survey question as it was originally worded is shown across the top of your screen. The respondents were asked to describe their marriage using three possible responses: "very happy," "pretty happy," or "not too happy." Looking at the pie chart we see that nearly two-thirds (63.2 percent) of the respondents indicate that they are very happily married and 33.8 percent say they are pretty happily married. Only 3 percent of those surveyed indicate that they are not too happily married.

Are marriages in our society really that happy? Perhaps, but researchers have found that even people who are not happily married may be reluctant to say so in an interview with a social researcher. Why? Some people may not be forthcoming because they are concerned with the way they appear to others—happy marriages are seen as an indicator of social and personal stability. Others may not even want to admit interpersonal failures to themselves. By the time some people get around to acknowledging the poor quality of their marital relationships, they will have already separated or divorced—in which case they would no longer be *unhappily married*. Thus, it is very difficult to locate those who identify them-

141

selves as unhappily married. As you go through this chapter, keep in mind the limitations of the variables used to measure the quality of marital relationships.

Even using this rough measure of marital happiness, we can observe some interesting differences. For example, look at marital happiness across the family life cycle.

Data File: **GSS**
➤ Task: **Cross-tabulation**
➤ Row Variable: **74) HAPP.MARR.?**
➤ Column Variable: **11) FAM.STAGE**
➤ View: **Tables**
➤ Display: **Column %**

HAP.MARR.?   by   FAM.STAGE
Cramer's V: 0.066 **

		FAM.STAGE				
		NO KIDS	WITH KIDS	EMPTY NEST	Missing	TOTAL
HAP MARR.?	VERY HAPPY	177	742	653	72	1572
		72.5%	59.6%	66.0%		63.4%
	PRET.HAPPY	59	468	306	46	833
		24.2%	37.6%	30.9%		33.6%
	NOT TOO	8	36	31	3	75
		3.3%	2.9%	3.1%		3.0%
	Missing	1	11	7	3029	3048
	TOTAL	244	1246	990	3150	2480
		100.0%	100.0%	100.0%		

What we see is a good example of a *curvilinear relationship*—one that is high at both ends and low in the middle. In this case, marital happiness is highest for those who have not had any children (72.5 percent) and those whose children are grown (66 percent), and lowest for those who have children still living at home (59.6 percent). Although the so-called empty-nesters are not quite as happy as those in the earliest stage of marriage, it is clearly those who are in the childrearing stage who have the lowest levels of marital happiness. Raising children requires time and energy, which means that there is less of both of these resources to devote to the marriage relationship.

Because women are more likely than men to spend much of their day with their children, we can hypothesize that **being in the child-rearing years will have more of an impact on marital happiness for women than for men**. To examine this relationship, we'll use 25) SEX as a control variable. This will produce one table that includes only males and another table that includes only females.

Data File: **GSS**
Task: **Cross-tabulation**
Row Variable: **74) HAPP.MARR.?**
Column Variable: **11) FAM.STAGE**
➤ Control Variable: **25) SEX**
➤ View: **Tables: MALE**
➤ Display: **Column %**

HAP.MARR.?   by   FAM.STAGE
Controls: SEX: MALE
Cramer's V: 0.105 **

		FAM.STAGE				
		NO KIDS	WITH KIDS	EMPTY NEST	Missing	TOTAL
HAP MARR.?	VERY HAPPY	71	332	334	35	737
		71.0%	57.6%	71.4%		64.4%
	PRET.HAPPY	25	228	125	24	378
		25.0%	39.6%	26.7%		33.0%
	NOT TOO	4	16	9	2	29
		4.0%	2.8%	1.9%		2.5%
	Missing	0	2	3	1251	1256
	TOTAL	100	576	468	1312	1144
		100.0%	100.0%	100.0%		

As indicated by the "Males" label at the top of the screen, the first table that appears includes only those responses for males. The results appear to be similar to those in the preceding table in which both men and women were included, except that the marital happiness for those in the empty-nest stage is equal with the levels of marital happiness in the first stage of marriage. These results too are statistically significant (V = 0.105**). Men who have children living in their household are less likely to describe their marriage as very happy.

Marriage and Family

Now let's look at the table for females:

Data File: **GSS**
Task: **Cross-tabulation**
Row Variable: **74) HAPP.MARR.?**
Column Variable: **11) FAM.STAGE**
Control Variable: **25) SEX**
➤ View: **Tables: FEMALE**
➤ Display: **Column %**

HAP.MARR.?　by　FAM.STAGE
Controls:　SEX: FEMALE
Cramer's V: 0.061 *

		FAM.STAGE				
		NO KIDS	WITH KIDS	EMPTY NEST	Missing	TOTAL
HAP.MARR.?	VERY HAPPY	106	410	319	37	835
		73.6%	61.2%	61.1%		62.5%
	PRET.HAPPY	34	240	181	22	455
		23.6%	35.8%	34.7%		34.1%
	NOT TOO	4	20	22	1	46
		2.8%	3.0%	4.2%		3.4%
	Missing	1	9	4	1778	1792
	TOTAL	144	670	522	1838	1336
		100.0%	100.0%	100.0%		

Click the appropriate button at the bottom of the task bar to look at the second (or "next") partial table for 25) SEX. This table includes only females.

Here we see that 73.6 percent of women who have not had any children say they are very happy with their marriage compared to 61.2 percent of those with children at home and 61.1 percent of those whose children are grown. The strength of the relationship is weaker than that for men because those in the empty-nest stage are almost identical with those in the child-rearing stage rather than those in the first stage of marriage. So, children detract from marital happiness for both men and women, but the effect for women continues even beyond the child-rearing years.

Do you think couples are more or less happily married today than they were in the past? During the past few decades, there has been a trend toward smaller families as many couples choose to have fewer or no children. Given that children tend to decrease marital happiness, are people today more happily married than they were in the 1970s?

➤ Data File: **TRENDS**
➤ Task: **Historical Trends**
➤ Variable: **18) HAP.MAR.?%**

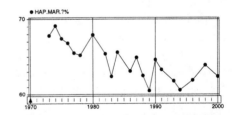

Percentage of respondents who say they are very happily married

Despite the trends toward smaller families, the percentage of people describing their marriages as very happy has declined since 1972. The range is fairly small, from a high of 68 percent in 1973 to a low of 61 percent in 1989, but there is clearly a downward trend overall. Influences other than family size must be affecting marital happiness.

Other factors that may influence marital happiness are the expectations people have going into marriage. For example, some people believe that husbands and wives should share work and family duties, whereas others believe it is better if spouses have separate roles with men being the providers and women taking care of the house and children. Let's see if these gender role preferences affect marital happiness.

➤ *Data File:* **GSS**
➤ *Task:* **Cross-tabulation**
➤ *Row Variable:* **74) HAP.MARR.?**
➤ *Column Variable:* **66) WIFE@HOME**
➤ *View:* **Tables**
➤ *Display:* **Column %**

HAP.MARR.?   by   WIFE@HOME
Cramer's V: 0.016

| | | WIFE@HOME | | | |
		AGREE	DISAGREE	Missing	TOTAL
	VERY HAPPY	433	612	599	1045
		62.8%	61.4%		61.9%
	PRET.HAPPY	236	356	287	592
		34.2%	35.7%		35.1%
	NOT TOO	21	29	28	50
		3.0%	2.9%		3.0%
	Missing	684	1251	1113	3048
	TOTAL	690	997	2027	1687
		100.0%	100.0%		

Among those who agree that it is better for the husband to be the provider and the wife to take care of the home, 62.8 percent describe their marriage as very happy compared to 61.4 percent who say they disagree with this statement. So, gender role attitudes are not very good predictors of marital happiness. But what about actual role enactment? Is there a difference in marital happiness between couples where both spouses are employed compared to those who have only one wage-earner?

Data File: **GSS**
Task: **Cross-tabulation**
Row Variable: **74) HAP.MARR.?**
➤ *Column Variable:* **31) FAM.WORK**
➤ *View:* **Tables**
➤ *Display:* **Column %**

HAP.MARR.?   by   FAM.WORK
Cramer's V: 0.036

| | | FAM.WORK | | | | |
		1 FULLTIME	1 FT/1 PT	2 FULLTIME	Missing	TOTAL
	VERY HAPPY	289	230	627	498	1146
		64.1%	64.2%	63.1%		63.6%
	PRET.HAPPY	152	114	346	267	612
		33.7%	31.8%	34.8%		34.0%
	NOT TOO	10	14	20	34	44
		2.2%	3.9%	2.0%		2.4%
	Missing	1	2	12	3033	3048
	TOTAL	451	358	993	3832	1802
		100.0%	100.0%	100.0%		

No, there is not a large difference in this table either. Couples where only one spouse is employed (64.1 percent) are equally likely to describe their marriage as very happy as those where both spouses are employed—either one full-time and one part-time (64.2 percent) or both full-time (63.1 percent). Just as gender role attitudes are not a good predictor of marital happiness, whether or not both spouses are employed does not impact a person's sense of marital happiness. Marriage appears to be a resilient and flexible institution that can flourish in a variety of forms.

The expectations people hold for marriage are largely under the control of the individual. Although it may not be easy, people may change their beliefs and attitudes about marriage. However, some of the circumstances that confront married couples are beyond their control but may still influence their marital happiness. For example, what effect does a person's health have on his or her marital happiness?

Data File: **GSS**
Task: **Cross-tabulation**
Row Variable: **74) HAP.MARR.?**
➤ Column Variable: **76) HEALTH**
➤ View: **Tables**
➤ Display: **Column %**

HAP.MARR.?   by   HEALTH
Cramer's V: 0.126 **

		HEALTH				
		EXCELLENT	GOOD	FAIR/POOR	Missing	TOTAL
HAP.MARR.?	VERY HAPPY	595	679	226	144	1500
		73.6%	60.0%	53.2%		63.5%
	PRET.HAPPY	204	415	171	89	790
		25.2%	36.7%	40.2%		33.4%
	NOT TOO	9	37	28	4	74
		1.1%	3.3%	6.6%		3.1%
	Missing	768	1317	696	267	3048
	TOTAL	808	1131	425	504	2364
		100.0%	100.0%	100.0%		

Reading across this table, we see that health does influence marital happiness. People who describe their health as excellent (73.6 percent) are the most happily married, while those who describe their health as fair or poor (53.2 percent) are the least happily married. Of course, it could be that a poor marriage itself takes a toll on a person's health. But it is also likely that the physical, emotional, and financial strains that can accompany poor health can decrease a person's marital happiness.

It might seem that marital happiness or dissatisfaction is unique to each couple—one couple is happy in their marriage, but another couple is quite unhappy in theirs. But, as you've seen countless times throughout this book, there are substantial regional variations to such things as divorce, birth rates, religion—many things that might initially seem to be based on personal decisions of individuals or couples. Personal happiness is no different. There is great variation across the nations of the world when it comes to family happiness or even general happiness in life. Let's switch to the GLOBAL data file and examine the general level of happiness people have in their lives. The following survey question was asked to respondents in approximately 40 countries: "Taking all things together, would you say you are very happy, quite happy, not very happy, or not at all happy?" The following variable indicates the percentage of people who say they are *very happy*.

➤ Data File: **GLOBAL**
➤ Task: **Mapping**
➤ Variable 1: **49) VERY HAPPY**
➤ View: **Map**

VERY HAPPY -- PERCENT WHO SAY THEY ARE VERY HAPPY (WVS)

Although this map has a lot of missing data, it is still clear that the nations of Eastern Europe (especially those that were formerly part of the Soviet Union) have the smallest number of happy people.

VERY HAPPY: Percent who say they are very happy

Data File: **GLOBAL**
Task: **Mapping**
Variable 1: **63) VERY HAPPY**
➤ View: **List: Rank**

RANK	CASE NAME	VALUE
1	Netherlands	47
2	Ireland	44
3	Denmark	43
4	Iceland	41
4	Sweden	41
6	Nigeria	40
6	Belgium	40
6	United States	40
9	United Kingdom	38
10	Switzerland	36

Nearly half of the respondents in the Netherlands (47 percent) indicate they are very happy. People in Ireland (44 percent), Denmark (43 percent), and Iceland (41 percent) are also on the cheery side. If you scroll to the bottom of the list, however, the picture looks downright dreary. Only 4 percent of people in Lithuania say they are very happy, followed by 3 percent of people in Estonia. Latvia is about as far down on this scale as you can get, having only 2 percent of people indicating they are very happy. Note that the United States is ranked sixth with a value of 40 percent.

A survey question relating specifically to happiness in one's family life was also asked: "Overall, how satisfied or dissatisfied are you with your home life?" Here is the percentage who say they are *very satisfied* with their home life.

HOME LIFE?: Percent who said they were "very satisfied" with their home life

Data File: **GLOBAL**
Task: **Mapping**
➤ Variable 1: **41) HOME LIFE?**
➤ View: **List: Rank**

RANK	CASE NAME	VALUE
1	Poland	65
2	Denmark	64
3	Switzerland	61
3	Ireland	61
5	Chile	58
6	United States	57
7	Canada	56
8	Brazil	55
8	Sweden	55
10	Finland	52

Here we see that 65 percent of people in Poland indicate they are very satisfied with their home life, followed by Denmark (64 percent) and Switzerland (61 percent). The United States is again ranked sixth with 57 percent of respondents indicating that they are very satisfied with their home life. In fact, if you examine the nations ranked highest on both variables, you'll see that about half of the same nations are located in the top ten.

Another indicator of family happiness might be the importance that one places on his or her family. Let's examine the results from the following survey question: "How important is family in your life?"

FAMILY IMP: Percent who say the family is very important in their lives

RANK	CASE NAME	VALUE
1	Nigeria	94
2	South Korea	93
2	United States	93
4	Canada	92
5	Iceland	91
5	Brazil	91
5	Ireland	91
5	Argentina	91
5	Poland	91
10	South Africa	90

Data File: **GLOBAL**
Task: **Mapping**
➤ Variable 1: **44) FAMILY IMP**
➤ View: **List: Rank**

Perhaps this measure is not as similar to the previous two variables as we thought. The ten highest ranked nations have at least 90 percent of survey respondents indicating that their family is a very important part of their lives. Even people from nations at the bottom of the list (China and Portugal are tied at 62 percent) indicate that family life is very important in their lives. Back at the top of this list, you'll see some familiar names, including the United States (93 percent), Iceland, Ireland, and Poland (the latter three are tied at 91 percent). Since the same nations keep appearing at the top of these lists, it seems highly likely that these three variables (VERY HAPPY, HOME LIFE?, and FAMILY IMP) will be positively correlated. Let's see.

Data File: **GLOBAL**
➤ Task: **Correlation**
➤ Variables: **44) FAMILY IMP**
41) HOME LIFE?
49) VERY HAPPY

Correlation Coefficients
PAIRWISE deletion (1-tailed test)   Significance Levels: ** = .01, * = .05

	FAMILY IMP	HOME LIFE?	VERY HAPPY
FAMILY IMP	1.000 (41)	0.537 ** (41)	0.467 ** (41)
HOME LIFE?	0.537 ** (41)	1.000 (41)	0.633 ** (41)
VERY HAPPY	0.467 ** (41)	0.633 ** (41)	1.000 (41)

Indeed, there are strong relationships between these variables and all are statistically significant. In the worksheet section that follows, you will see what other factors are related to these variables. You will also return to the marital happiness variable in the GSS to see how it is related to other social, economic, and familial factors.

WORKSHEET

CHAPTER 9

NAME:

COURSE:

DATE:

REVIEW QUESTIONS

Based on the first part of this chapter, answer True or False to the following items:

The majority of people in the United States say they are very happily married.	T	F
Children affect marital happiness for women but not men.	T	F
Couples with children living at home are the most likely to say they are very happily married.	T	F
People who disagree that it is better for the husband to be the provider while the wife takes care of the house are more likely to describe their marriage as very happy.	T	F
Single-career and dual-career couples have about the same levels of marital happiness.	T	F
The percentage of people in the United States who describe their marriage as very happy has been declining since 1972.	T	F
Compared to other nations, people in the United States are very happy in their home life.	T	F

EXPLORIT QUESTIONS

1. Let's continue with where we left off in the preliminary section of this chapter. We have already used the GLOBAL file to compare nations in terms of the importance people place on their family and whether people are satisfied with family life and their personal lives. Once again, the three survey questions that were asked are

    49) VERY HAPPY—*Taking all things together, would you say you are very happy, quite happy, not very happy, or not at all happy?*

    41) HOME LIFE?—*Overall, how satisfied or dissatisfied are you with your home life?*

    44) FAMILY IMP—*How important is family in your life?*

    a. In your opinion, which of these three questions is the best indicator of marital happiness? _____

b. Why do you think the variable you selected is the best indicator? Why are the other two questions not as good for assessing marital happiness?

2. Let's see how national attitudes on these indicators are related to religious practices within nations.

> *Data File:* **GLOBAL**
> *Task:* **Correlation**
> *Variables:* **49) VERY HAPPY**
> **41) HOME LIFE?**
> **44) FAMILY IMP**
> **59) CH.ATTEND** *(Percentage who attend religious services once a month or more)*
> **60) GOD IMPORT** *(Percentage saying God is important in their lives)*
> **61) PRAY?** *(Percentage who pray at least sometimes)*

a. Print this correlation matrix and attach it to your assignment.

b. Nations with high percentages of people who attend religious services are more likely to have people who indicate that (circle all responses that are supported by the results)
   1. they are very happy with their life.
   2. they are very satisfied with their home life.
   3. family is an important part of their life.

c. Nations with high percentages of people who indicate that God is important in their life are more likely to have people who indicate that (circle all responses that are supported by the results)
   1. they are very happy with their life.
   2. they are very satisfied with their home life.
   3. family is an important part of their life.

d. Nations with high percentages of people who pray are more likely to have people who indicate that (circle all responses that are supported by the results)
   1. they are very happy with their life.
   2. they are very satisfied with their home life.
   3. family is an important part of their life.

e. Overall, are religious practices within nations more related to attitudes people have toward the importance of the family, or are they more strongly related to general happiness and family happiness? (circle one)
   1. more strongly related to the importance of family
   2. more strongly related to general happiness and family happiness
   3. equally related to the importance of family, general happiness, and family happiness

f. In general, do religious practices within nations seem to be strongly related to the
variable you selected in Question 1a?   Yes   No

3. Religious practices are different for each type of religion. For example, prayer in one religion may
consist of a brief recital of a sentence or two, whereas in other religions it may be viewed as a highly
personal discussion with God. Repeat the previous analysis, except this time compare Muslim countries with those that are predominantly Christian.

> Data File: **GLOBAL**
> Task: **Correlation**
> ➤ Variables: **49) VERY HAPPY**
> **41) HOME LIFE?**
> **44) FAMILY IMP**
> **53) %MUSLIM**
> **54) %CHRISTIAN**

a. In a few sentences, explain why you agree or disagree with the following statement: "Both the
%MUSLIM and %CHRISTIAN variables are good predictors of the variable HOME LIFE?. That is,
if we know the percentage of people in a country who are either Islamic or Christian, we can predict with a good degree of accuracy the percentage of people in the nation who are very satisfied
with their home life."

b. Is %MUSLIM or %CHRISTIAN a better predictor of the variable FAMILY IMP?   _____

c. Are %MUSLIM and %CHRISTIAN good predictors of VERY HAPPY?   Yes   No

4. In the preliminary part of this chapter, we used the GSS data to examine factors that might be related
to marital satisfaction. Let's pursue this analysis a bit further. Because those who have been married
more than once will have learned from their previous marriages what they desire in a mate, we'll predict that **those who are divorced and remarried will be more likely to say they are very happy
with their current marriages**.

> ➤ Data File: **GSS**
> ➤ Task: **Cross-tabulation**
> ➤ Row Variable: **74) HAP.MARR.?**
> ➤ Column Variable: **9) REMARRIAGE**
> ➤ View: **Tables**
> ➤ Display: **Column %**

a. Fill in the percentaged results for the *top* row of the table.

	1 MARRIAGE	REMARRIAGE
VERY HAPPY	_____%	_____%

b. What is the value of V?   V = _____

c. Is the hypothesis supported?   Yes   No

5. Does a person's family income impact his or her marital happiness?

> *Data File:* **GSS**
> *Task:* **Cross-tabulation**
> *Row Variable:* **74) HAP.MARR.?**
> ➤ *Column Variable:* **28) INCOME**
> ➤ *View:* **Tables**
> ➤ *Display:* **Column %**

a. Fill in the percentaged results for the *top* row of the table.

	LOW	MIDDLE	HIGH
VERY HAPPY	_____%	_____%	_____%

b. What is the value of V?   V = _____

c. Is V statistically significant?   Yes   No

d. Based on these results, in which income level are people *most* likely to say they are very happy? (circle one)
     1. low
     2. middle
     3. high

e. In which social classes are people *least* likely to say they are very happily married? (circle one)
     1. low
     2. middle
     3. high

f. How would you explain these results?

6. Does a change in income pull families together or push them apart? The GSS asked the following question: "During the last few years, has your financial situation been getting better, worse, or has it stayed the same?"

> Data File: **GSS**
> Task: **Cross-tabulation**
> Row Variable: **74) HAP.MARR.?**
> ➤ Column Variable: **35) CHANGE $?**
> ➤ View: **Tables**
> ➤ Display: **Column %**

a. Fill in the percentaged results for the *top* row of the table.

	BETTER	WORSE	THE SAME
VERY HAPPY	_____%	_____%	_____%

b. What is the value of V?   V = _____

c. Is V statistically significant?   Yes   No

d. People whose economic situation has changed for the worse are (circle one)
    1. less likely to say they are very happily married.
    2. more likely to say they are very happily married.
    3. no more or less likely to say they are very happily married.

e. Based on the results of the previous two tables, it appears that economic hardship (circle one)
    1. places a strain on marital relationships and decreases marital happiness.
    2. pulls most couples together and increases marital happiness.
    3. has no impact one way or the other on marital happiness.

7. Are race and marital satisfaction related?

> Data File: **GSS**
> Task: **Cross-tabulation**
> Row Variable: **74) HAP.MARR.?**
> ➤ Column Variable: **26) RACE**
> ➤ View: **Tables**
> ➤ Display: **Column %**

a. Fill in the percentaged results for the *top* row of the table.

	WHITE	BLACK
VERY HAPPY	_____%	_____%

b. What is the value of V?

$V =$ _____

c. Is V statistically significant?

Yes   No

8. Construct the previous table again, only this time exclude those who are in the highest income category.

> Data File: **GSS**
> Task: **Cross-tabulation**
> Row Variable: **74) HAP.MARR.?**
> Column Variable: **26) RACE**
> ➤ Subset Variable: **28) INCOME**
> ➤ Subset Category: **Exclude 3) HIGH**
> ➤ View: **Tables**
> ➤ Display: **Column %**

a. Fill in the percentaged results for the *top* row of the table.

	WHITE	BLACK
VERY HAPPY	_____%	_____%

b. What is the value of V?

$V =$ _____

c. Is V statistically significant?

Yes   No

d. Summarize what happens to the relationship between race and marital happiness once income is controlled.

9. Are church attendance and marital satisfaction related?

> Data File: **GSS**
> Task: **Cross-tabulation**
> Row Variable: **74) HAP.MARR.?**
> ➤ Column Variable: **50) ATTEND**
> ➤ View: **Tables**
> ➤ Display: **Column %**

Remember to remove the subset variable before continuing.

a. Fill in the percentaged results for the *top* row of the table.

	NEVER	MONTH/YRLY	WEEKLY
VERY HAPPY	_____%	_____%	_____%

b. What is the value of V?   V = _____

c. Is V statistically significant?   Yes   No

d. In a few sentences, indicate what you think might explain these results.

10. Finally, let's see if there are regional differences in marital satisfaction across the United States.

> *Data File:* **GSS**
> *Task:* **Cross-tabulation**
> *Row Variable:* **74) HAP.MARR.?**
> ➤ *Column Variable:* **37) REGION**
> ➤ *View:* **Tables**
> ➤ *Display:* **Column %**

Remember to remove the subset variable before continuing.

a. Fill in the percentaged results for the *top* row of the table.

	EAST	MIDWEST	SOUTH	WEST
VERY HAPPY	_____%	_____%	_____%	_____%

b. What is the value of V?   V = _____

c. Is V statistically significant?   Yes   No

d. People from which region have the highest levels of marital happiness?   _____

11. The South is often referred to as the *Bible Belt*, so let's see what happens if we control for the influence of religion on the relationship between 37) REGION and 74) HAP.MARR.?

> *Data File:* **GSS**
> *Task:* **Cross-tabulation**
> *Row Variable:* **74) HAP.MARR.?**
> *Column Variable:* **37) REGION**
> ➤ *Subset Variable:* **50) ATTEND**
> ➤ *Subset Category:* **Include 1) WEEKLY**
> ➤ *View:* **Tables**
> ➤ *Display:* **Column %**

a. Fill in the percentaged results for the *top* row of the table.

	EAST	MIDWEST	SOUTH	WEST
VERY HAPPY	_____%	_____%	_____%	_____%

b. What is the value of V?   V = _____

c. Is V statistically significant?   Yes   No

d. Based upon these results, church attendance has more of an influence on marital happiness than does the region in which someone lives.   T   F

12. Many people believe that being happy personally is the key to finding marital happiness. Let's see if personal happiness is indeed related to marital happiness.

> *Data File:* **GSS**
> *Task:* **Cross-tabulation**
> *Row Variable:* **74) HAP.MARR.?**
> ➤ *Column Variable:* **73) HAPPY?**
> ➤ *View:* **Tables**
> ➤ *Display:* **Column %**

Remember to remove the subset variable before continuing.

a. Fill in the percentaged results for the *top* row of the table.

	VERY HAPPY	PRET.HAPPY	NOT TOO
VERY HAPPY	_____%	_____%	_____%

b. Is V statistically significant?   Yes   No

c. Personal happiness and marital happiness are related to each other.   T   F

d. Which of these variables do you think comes first in time—marital happiness or personal happiness? Explain your answer.

CHAPTER 10

FERTILITY AND FAMILY LIFE

Tasks: Univariate, Cross-tabulation, Historical Trends, Mapping, Correlation, Scatterplot
Data Files: GSS, TRENDS, GLOBAL

Among the most important decisions a couple must make is whether or not to have children. If a couple decides to have children, then the next question is how many. Their hope for a certain family size may or may not be realized. Some couples end up with more children than they had planned, others with fewer. In either case, the impact of children on the marriage relationship is still an important topic to explore. In this chapter, we will explore some variations in family size, then examine the impact of children on family life

Let's begin by using the GLOBAL file to examine fertility rates across the nations of the world. The variable 5) FERTILITY indicates the number of children born to an average woman in her lifetime.

> *Data File:* **GLOBAL**
> *Task:* **Mapping**
> *Variable 1:* **5) FERTILITY**
> *View:* **Map**
> *Display:* **Legend**

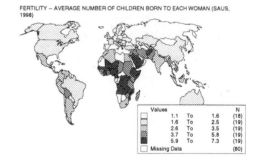

FERTILITY -- AVERAGE NUMBER OF CHILDREN BORN TO EACH WOMAN (SAUS, 1998)

Values			N
1.1	To	1.6	(18)
1.6	To	2.5	(19)
2.6	To	3.5	(19)
3.7	To	5.8	(19)
5.9	To	7.3	(19)
Missing Data			(80)

You'll rarely see a map with such a clear regional pattern. It is immediately evident that fertility rates are highest in Africa and the Middle Eastern regions. The rates are lowest throughout all of Europe and North America. If you examine the map legend, you see that the darkest shading represents the countries in which the average woman has 5.9 to 7.3 children during the course of her life. Women who live in countries having the lightest shading have an average of 1.1 to 1.6 children in a lifetime—that's over 5 fewer children per woman! Now look at a ranked list for these nations.

FERTILITY: Average number of children born to each woman

Data File: **GLOBAL**
Task: **Mapping**
Variable 1: **5) FERTILITY**
➤ *View:* **List: Rank**

RANK	CASE NAME	VALUE
1	Niger	7.3
2	Yemen	7.1
3	Uganda	7.1
4	Mali	7.0
5	Somalia	7.0
6	Nigeria	6.9
7	Ethiopia	6.9
8	Burkina Faso	6.6
9	Congo, Dem. Republic	6.5
10	Benin	6.5

Niger ranks at the top of the list with a fertility rate of 7.3. It is followed by dozens of African and Middle Eastern nations. At the bottom of the list are many European nations, including Bulgaria (1.1), the Czech Republic (1.2), Romania (1.2), Italy (1.2), and Spain (1.2). The United States ranks 65th with 2.1 children being born to the average woman in her lifetime.

Keep the fertility variable selected so that you can compare it to the map for infant mortality.

Data File: **GLOBAL**
Task: **Mapping**
Variable 1: **5) FERTILITY**
➤ *Variable 2:* **7) INF. MORTL**
➤ *Views:* **Map**

FERTILITY -- AVERAGE NUMBER OF CHILDREN BORN TO EACH WOMAN (SAUS, 1998)

r = 0.847**

INF. MORTL -- NUMBER OF INFANT DEATHS PER 1,000 BIRTHS (TWF, 1997)

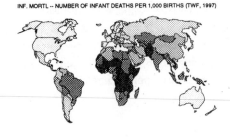

Wait, are these the same maps? No, but they certainly look similar. The correlation between fertility and infant mortality rates is extremely high (r = 0.847**). So it seems that high rates of birth—and infant deaths—might occur in poor countries that lack health care services.

Marriage and Family

Let's pursue this idea a bit further using the CORRELATION task. We'll reexamine the correlation between fertility and infant mortality, but add two more variables: MOM MORTAL (maternal mortality rates) and $ PER CAP (annual gross national product per capita).

Data File: **GLOBAL**
➤ Task: **Correlation**
➤ Variables: **5) FERTILITY**
   7) INF. MORTL
   8) MOM MORTAL
   22) $ PER CAP

Correlation Coefficients
PAIRWISE deletion (1-tailed test)   Significance Levels: ** = .01, * = .05

	FERTILITY	INF. MORTL	MOM MORTAL	$ PER CAP
FERTILITY	1.000 (94)	0.847 ** (94)	0.820 ** (88)	-0.593 ** (94)
INF. MORTL	0.847 ** (94)	1.000 (174)	0.828 ** (140)	-0.671 ** (173)
MOM MORTAL	0.820 ** (88)	0.828 ** (140)	1.000 (140)	-0.536 ** (140)
$ PER CAP	-0.593 ** (94)	-0.671 ** (173)	-0.536 ** (140)	1.000 (173)

There is no question about it. Countries that have high fertility rates have high infant mortality rates, high maternal mortality rates (i.e., deaths by mothers in childbirth), and low levels of gross national product per capita. The poorest countries have the highest fertility rates.

Is there a direct relationship between the percentage of women who use contraceptives and the fertility rate within nations? It seems like there should be an inverse (negative) relationship.

Data File: **GLOBAL**
➤ Task: **Scatterplot**
➤ Dependent Variable: **5) FERTILITY**
➤ Independent Variable: **9) CONTRACEPT**
➤ Display: **Reg. Line**

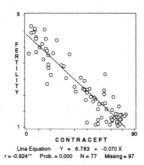

Line Equation   Y = 6.783 + -0.070 X
r = -0.924**   Prob. = 0.000   N = 77   Missing = 97

Perhaps the only surprise here is the strength of the negative correlation. Pearson's r is a whopping –0.924**. Clearly, the use of contraceptives is strongly related to reduced fertility rates.

As we saw earlier, the average woman in the United States has 2.1 children during her lifetime. Let's take a closer look at fertility rates in the United States and try to gain a better understanding of why some women have more children than others. We'll begin by looking at the historical trend in fertility rates during the twentieth century.

➤ Data File: **TRENDS**
➤ Task: **Historical Trends**
➤ Variable: **14) BIRTH RATE**

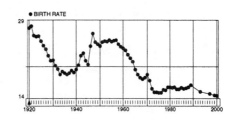

Estimated births per 1,000 population

One of the first things you will probably notice is the hump in the middle of this graph that stretches from the mid-1940s to the early 1960s—this is the baby boom that people often talk about. Prior to this period of time, fertility rates had been declining as the strain of the Great Depression made it difficult for people to afford large families. But, with the affluence of the post–World War II economy and the accompanying surge in pro-family values, came a rapid increase in the birth rate. However, during the 1960s and 1970s, fertility rates dropped as sharply as they had increased earlier, and the trend toward lower birth rates resumed.

What has caused this decline in the fertility rate? At least part of the answer lies in alternatives to motherhood that came with increased industrialization.

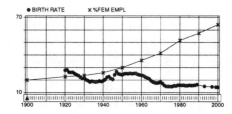

Data File:	**TRENDS**
Task:	**Historical Trends**
➤ *Variables:*	**14) BIRTH RATE**
	26) %FEM EMPL

Estimated births per 1,000 population
Percent of females employed full-time outside the home

As you can see, the female employment rate rises sharply at roughly the same time when birth rates start their decline following the baby boom. Prior to industrialization women could combine work and family at home, but that is not possible when production shifts to the factory. Thus, as female employment outside the home increases—the birth rate decreases.

Let's shift to the General Social Survey to explore these fertility trends at an individual level.

➤ *Data File:*	**GSS**
➤ *Task:*	**Univariate**
➤ *Primary Variable:*	**13) # CHILDREN**
➤ *View:*	**Pie**

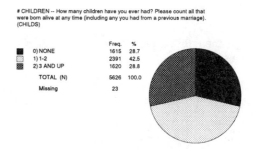

This figure shows how many children survey respondents from the GSS have had over their lifetime. Among those surveyed, 42.5 percent have had one or two children. The rest are evenly divided between those who have had three or more children and those who have had none. Let's see how these percentages compare to what respondents believe is the ideal family size.

Marriage and Family

Data File: **GSS**

Task: **Univariate**

➤ Primary Variable: **14) IDEAL#KIDS**

➤ View: **Pie**

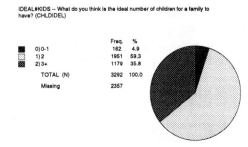

IDEAL#KIDS -- What do you think is the ideal number of children for a family to have? (CHLDIDEL)

		Freq.	%
■	0) 0-1	162	4.9
▦	1) 2	1951	59.3
▨	2) 3+	1179	35.8
	TOTAL (N)	3292	100.0
	Missing	2357	

Over half (59.3 percent) of those surveyed believe that the ideal number of children to have is two. Given the decline in the birth rate, what may surprise you is that only 4.9 percent believe that it is ideal to have just one child or no children at all. Most of those who vary from the norm of believing that having two children is ideal seem to desire more children rather than fewer. Despite the declining birth rate in the United States, we are still a very pro-natalistic culture.

We have already seen that the fertility rate is declining in the United States, so let's see if we can find evidence of this trend in the General Social Survey. One way we can do this is by dividing the sample into age "cohorts" and then comparing the younger generations with the older generations.

Data File: **GSS**

➤ Task: **Cross-tabulation**

➤ Row Variable: **13) # CHILDREN**

➤ Column Variable: **24) AGE**

➤ View: **Tables**

➤ Display: **Column %**

CHILDREN   by   AGE
Cramer's V: 0.311 **

		AGE				
		<30	30-49	50 AND UP	Missing	TOTAL
# CHILDREN	NONE	682	638	291	4	1611
		64.4%	25.5%	14.2%		28.7%
	1-2	302	1245	837	7	2384
		28.5%	49.7%	40.8%		42.5%
	3 AND UP	75	620	924	1	1619
		7.1%	24.8%	45.0%		28.8%
	Missing	4	8	11	0	23
	TOTAL	1059	2503	2052	12	5614
		100.0%	100.0%	100.0%		

As expected, the older groups tend to have more children. The most distinct dividing line is at age 50. Look at the row labeled "3 AND UP." The percentage of respondents over age 50 with three or more children is nearly double that of previous generations. Of course, these figures may be deceiving, because many of the younger respondents are likely to have more children in the future. However, people who are currently age 30–49 are not likely to substantially increase the number of children they have over the coming decade to match those obtained by the current 50–64 age cohort.

So, is the average number of children decreasing because some couples are deciding to have no children at all (and, hence, dragging down the average), or are the majority of married couples simply having fewer children? The trend line below shows the percentage of married (or previously married) respondents who have no children. The time span is 1972 to 2000.

➤ *Data File:* **TRENDS**
    ➤ *Task:* **Historical Trends**
➤ *Variable:* **16) NO KIDS%**

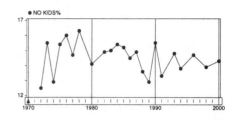

Percentage of married, or previously married, respondents who do not have any children

The percentage of married couples who have no children has actually changed very little since 1972. There is some fluctuation, but the range is only from about 12 percent to 16 percent. Ultimately, married couples in 2000 were only about 2 percent more likely than married couples in 1972 to be without children.

Now we'll look at the percentage of married (or previously married) respondents having three or more children. Again, the time span is 1972 to 2000.

Data File: **TRENDS**
    Task: **Historical Trends**
➤ *Variable:* **17) KIDS 3+%**

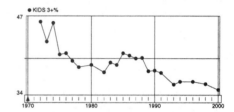

Percentage of married, or previously married, respondents with 3 or more children

Indeed, this seems to be one of the reasons for declining fertility rates in the United States. The percentage of the population having three or more children has declined over the last few decades. In 1972, nearly 47 percent of those surveyed had three or more children, compared to approximately 35 percent in 2000. Most of the decline takes place in the 1970s. The percentage then levels off in the 1980s before declining again in the 1990s.

The increasing effectiveness and availability of contraceptives has given individuals a considerable amount of control over their fertility. Of course, not every pregnancy is planned, and not everyone who plans to get pregnant is able to do so, but the freedom to make childbearing decisions has never been greater than it is now. But individuals do not make fertility-related decisions in a vacuum; such decisions are influenced by the social environment. For example, earlier we saw that the decline in the fertility rate in the United States corresponds with an increase in the female employment rate. Today, most married women, including those with children, are employed outside the home (this trend will be discussed in Chapter 12). When both parents are employed, there is less time available to devote to parenting, especially if both parents are employed full-time (dual-career families). Thus, two-income families are likely to have fewer children so that they can balance their work and family roles. Families in which one spouse is a full-time homemaker, on the other hand, will probably have more children, because there is more time to devote to parenting and because more of the homemaker's identity is derived from the parenting role. Dual-earner families, in which both spouses are employed, but at least one works only part-time, will fall in the middle. Therefore, the hypothesis is: **Married**

couples with only one employed spouse will have the most children, and couples in which both spouses have careers will have the fewest children.

> ➤ Data File: **GSS**
> ➤ Task: **Cross-tabulation**
> ➤ Row Variable: **13) # CHILDREN**
> ➤ Column Variable: **31) FAM.WORK**
> ➤ View: **Tables**
> ➤ Display: **Column %**

CHILDREN   by   FAM.WORK
Cramer's V: 0.125 **

		FAM.WORK				
		1 FULLTIME	1 FT/1 PT	2 FULLTIME	Missing	TOTAL
# CHILDREN	NONE	29	38	194	1354	261
		6.4%	10.6%	19.3%		14.4%
	1-2	253	216	555	1367	1024
		56.2%	60.0%	55.3%		56.5%
	3 AND UP	168	106	254	1092	528
		37.3%	29.4%	25.3%		29.1%
	Missing	2	0	2	19	23
	TOTAL	450	360	1003	3832	1813
		100.0%	100.0%	100.0%		

Examine the bottom row and you'll see that 37.3 percent of single-earner families have three or more children compared to 25.3 percent of families with both spouses having full-time jobs. In between these values are families in which one spouse works full-time and the other works part-time (29.4 percent). If you look across the top row, you will see that when both spouses work full-time they are more likely to have no children. The difference is statistically significant (V = .125**). In industrialized societies, where work usually takes place outside the home, married couples have reduced their family size to make parenting more manageable.

In addition to affecting the dynamics of everyday family life, industrialization creates a demand for an educated workforce. Today, most people receive at least some college training. For many young people, going to college will delay marriage and/or childbearing. The investment in a college degree may also lead to a greater commitment to the workplace after graduation—which, as we have already seen, limits family size. Either way, we would hypothesize that **those who have more education will have fewer children.**

> Data File: **GSS**
> Task: **Cross-tabulation**
> Row Variable: **13) # CHILDREN**
> ➤ Column Variable: **27) EDUCATION**
> ➤ View: **Tables**
> ➤ Display: **Column %**

CHILDREN   by   EDUCATION
Cramer's V: 0.132 **

		EDUCATION				
		NO HS GRAD	HS GRAD	COLL EDUC	Missing	TOTAL
# CHILDREN	NONE	187	383	1041	4	1611
		19.2%	23.0%	35.1%		28.7%
	1-2	381	756	1246	8	2383
		39.1%	45.3%	42.0%		42.5%
	3 AND UP	406	529	678	7	1613
		41.7%	31.7%	22.9%		28.8%
	Missing	2	6	13	2	23
	TOTAL	974	1668	2965	21	5607
		100.0%	100.0%	100.0%		

This hypothesis appears to be supported: the percentage of respondents having three or more children decreases as the education level increases. Forty-one percent of those who do not have a high school degree have three or more children compared to 31.7 percent of those with a high school degree and 22.9 percent of those who went to college. Of course, we can't say for sure which comes first—having children or dropping out of school. For some, the demands of having children probably led to dropping out of school. But, for others, not having educational goals to pursue may have led to having more children. Either way, there is an inverse relationship between large families and higher education.

Let's look for some other variations in family size. Do whites and African Americans differ in family size?

Data File: **GSS**
Task: **Cross-tabulation**
Row Variable: **13) # CHILDREN**
➤ Column Variable: **26) RACE**
➤ View: **Tables**
➤ Display: **Column %**

CHILDREN   by   RACE
Cramer's V: 0.099 **

		RACE			
		WHITE	BLACK	Missing	TOTAL
# CHILDREN	NONE	1330	165	120	1495
		29.8%	19.9%		28.3%
	1-2	1906	348	137	2254
		42.7%	42.0%		42.6%
	3 AND UP	1227	316	77	1543
		27.5%	38.1%		29.2%
	Missing	16	3	4	23
	TOTAL	4463	829	338	5292
		100.0%	100.0%		

Whites (29.8 percent) are more likely than African Americans (19.9 percent) to have no children, and African Americans (38.1 percent) are more likely than whites (27.5 percent) to have three or more children. The difference is statistically significant (V= 0.099**). Are these differences based on a difference in cultural values, or do they result from fewer educational and economic opportunities? We can partially answer this question by controlling for education.

Data File: **GSS**
Task: **Cross-tabulation**
Row Variable: **13) # CHILDREN**
Column Variable: **26) RACE**
➤ Subset Variable: **27) EDUCATION**
➤ Subset Category: **Include: 3) COLL EDUC**
➤ View: **Tables**
➤ Display: **Column %**

CHILDREN   by   RACE

Cramer's V: 0.086 **

		RACE			
		WHITE	BLACK	Missing	TOTAL
# CHILDREN	NONE	879	79	83	958
		36.0%	23.5%		34.5%
	1-2	1005	169	72	1174
		41.1%	50.3%		42.2%
	3 AND UP	560	88	30	648
		22.9%	26.2%		23.3%
	Missing	8	1	4	13
	TOTAL	2444	336	189	2780
		100.0%	100.0%		

When we look only at the subset of those who have at least some college education, the biggest difference is that whites are more likely to have no children. The gap between whites (22.9 percent) and blacks (26.2 percent) with three or more children is relatively small. Let's use this same education subgroup to see if, in fact, African Americans desire to have larger families.

Data File: **GSS**

Task: **Cross-tabulation**

➤ Row Variable: **14) IDEAL#KIDS**

➤ Column Variable: **26) RACE**

➤ Subset Variable: **27) EDUCATION**

➤ Subset Category: **Include: 3) Coll Educ**

➤ View: **Tables**

➤ Display: **Column %**

IDEAL#KIDS   by   RACE

Cramer's V: 0.091 **

		RACE			
		WHITE	BLACK	Missing	TOTAL
IDEAL#KIDS	0-1	69	11	4	80
		4.8%	5.5%		4.9%
	2	928	102	68	1030
		64.5%	51.3%		62.9%
	3+	442	86	43	528
		30.7%	43.2%		32.2%
	Missing	1013	138	74	1225
	TOTAL	1439	199	189	1638
		100.0%	100.0%		

The differences in this table are larger than the differences in the previous table. Among African Americans, 43.2 percent say the ideal number of children is three or more, compared to 30.7 percent of whites who feel the same way. There is a preference for larger families among African Americans (V = 0.091**).

Finally, look at another cultural variable that many assume to be related to family size—religion. Because the Catholic Church officially forbids the use of contraceptives, we would hypothesize that Catholics will be the most likely to have three or more children. (Be sure to delete the subset variable from the previous analysis before proceeding.)

Data File: **GSS**

Task: **Cross-tabulation**

➤ Row Variable: **13) # CHILDREN**

➤ Column Variable: **48) RELIGION**

➤ View: **Tables**

➤ Display: **Column %**

CHILDREN   by   RELIGION

Cramer's V: 0.128 **

		RELIGION						
		LIB. PROT.	CON.PROT.	CATHOLIC	JEWISH	NONE	Missing	TOTAL
# CHILDREN	NONE	287	346	409	38	349	186	1429
		25.1%	21.3%	29.7%	33.9%	44.3%		28.3%
	1-2	500	718	577	50	300	246	2145
		43.7%	44.3%	41.9%	44.6%	38.1%		42.5%
	3 AND UP	358	557	392	24	139	150	1470
		31.3%	34.4%	28.4%	21.4%	17.6%		29.1%
	Missing	3	3	6	1	6	4	23
	TOTAL	1145	1621	1378	112	788	586	5044
		100.0%	100.0%	100.0%	100.0%	100.0%		

Make sure to delete the subset variable 27) EDUCATION that was selected for the previous analysis. If you prefer, you can click the [Clear All] button to clear all previously selected variables.

The results do not support the hypothesis. Both liberal Protestants (31.3 percent) and conservative Protestants (34.4 percent) are somewhat more likely than Catholics (28.4 percent) to have three or more children. There is a difference, however, when one compares those who have a religious preference with those who do not. Respondents with no religious affiliation (17.6 percent) are the least likely to have three or more children (V = .128**).

So far we have seen that the decision of how many children to have is influenced by such things as the female employment rate, education, race, and religion. But the choice of how many children to have is not the only childbearing decision that has to be made—there also is the question of timing. What is the best age at which to have a child? Of course, there are many factors to consider when making this decision, but one thing to consider is how having children at different ages affects educational attainment and family income.

Data File: **GSS**
Task: **Cross-tabulation**
➤ Row Variable: **27) EDUCATION**
➤ Column Variable: **18) AGE KD BRN**
➤ View: **Tables**
➤ Display: **Column %**

EDUCATION   by   AGE KD BRN
Cramer's V: 0.214 **

		AGE KD BRN				
		<20	20-29	30 AND UP	Missing	TOTAL
E D U C A T I O N	NO HS GRAD	357	354	63	202	774
		37.7%	14.4%	11.7%		19.6%
	HS GRAD	328	826	118	402	1272
		34.7%	33.5%	21.9%		32.2%
	COLL EDUC	261	1283	359	1075	1903
		27.6%	52.1%	66.5%		48.2%
	Missing	6	5	4	6	21
	TOTAL	946	2463	540	1685	3949
		100.0%	100.0%	100.0%		

The age at which respondents had their first child is related to their level of education. Those who had a child while they were a teenager (37.7 percent) are about one-third as likely to have graduated from high school compared to those who waited until they were at least 20 (14.4 percent). Conversely, those who waited until at least age 30 (66.5 percent) are the most likely to have gone to college, compared to 52.1 percent for those age 20–29 and 27.6 percent for those under age 20 when their first child was born. The difference is statistically significant (V = .214**).

Do the educational consequences of having a child at a young age affect men or just women?

Data File: **GSS**
Task: **Cross-tabulation**
Row Variable: **27) EDUCATION**
Column Variable: **18) AGE KD BRN**
➤ Control Variable: **25) SEX**
➤ View: **Tables (MALE)**
➤ Display: **Column %**

EDUCATION   by   AGE KD BRN
Controls: SEX: MALE
Cramer's V: 0.168 **

		AGE KD BRN				
		<20	20-29	30 AND UP	Missing	TOTAL
E D U C A T I O N	NO HS GRAD	79	187	42	119	308
		39.1%	17.4%	13.9%		19.5%
	HS GRAD	65	339	62	215	466
		32.2%	31.6%	20.5%		29.5%
	COLL EDUC	58	547	198	541	803
		28.7%	51.0%	65.6%		50.9%
	Missing	1	1	3	4	9
	TOTAL	202	1073	302	879	1577
		100.0%	100.0%	100.0%		

Here we see the results just for men, and they are almost identical to the results in the previous table. The younger a male was when his first child was born, the lower his level of education (V = .168**). How does this compare to the results for women?

Data File: **GSS**
Task: **Cross-tabulation**
Row Variable: **27) EDUCATION**
Column Variable: **18) AGE KD BRN**
Control Variable: **25) SEX**
➤ View: **Tables (FEMALE)**
➤ Display: **Column %**

EDUCATION   by   AGE KD BRN
Controls: SEX: FEMALE
Cramer's V: 0.242 **

		AGE KD BRN				
		<20	20-29	30 AND UP	Missing	TOTAL
EDUCATION	NO HS GRAD	278	167	21	83	466
		37.4%	12.0%	8.8%		19.6%
	HS GRAD	263	487	56	187	806
		35.3%	35.0%	23.5%		34.0%
	COLL EDUC	203	736	161	534	1100
		27.3%	52.9%	67.6%		46.4%
	Missing	5	4	1	2	12
	TOTAL	744	1390	238	806	2372
		100.0%	100.0%	100.0%		

Click the appropriate button at the bottom of the task bar to look at the second (or next) partial table for 25) SEX.

As would be expected, the relationship between having a baby at a young age is even greater for women than it was for men (V = 0.242**).

We know that education and income are related, so let's test the hypothesis that **those who had their first child when they were older will have higher incomes.**

Data File: **GSS**
Task: **Cross-tabulation**
➤ Row Variable: **28) INCOME**
➤ Column Variable: **18) AGE KD BRN**
➤ View: **Tables**
➤ Display: **Column %**

INCOME   by   AGE KD BRN
Cramer's V: 0.189 **

		AGE KD BRN				
		<20	20-29	30 AND UP	Missing	TOTAL
INCOME	LOW	356	477	81	447	914
		42.7%	22.0%	17.3%		26.3%
	MIDDLE	331	888	139	631	1358
		39.7%	40.9%	29.7%		39.1%
	HIGH	146	805	248	410	1199
		17.5%	37.1%	53.0%		34.5%
	Missing	119	298	76	197	690
	TOTAL	833	2170	468	1685	3471
		100.0%	100.0%	100.0%		

Remember to remove the control variable before continuing.

As expected, those who waited until their thirties to have their first child are the least likely to be in the low income category (17.3 percent) and the most likely to be in the high income category (53 percent). The percentages are almost exactly reversed for those who had a child in their teens, with 42.7 percent being in the low income category and only 17.5 percent in the high income category. The hypothesis is supported; those who wait to have their first child tend to have higher incomes (V = 0.189**).

Because the timing of having children has a lasting impact on the parents, it is important that individuals make deliberate choices in this regard. The same can be said for the choice of how many children to have—it is a decision that has lasting consequences. In the worksheet section, you will have the opportunity to explore some of the ways family size impacts personal lifestyles along with exploring some other fertility-related issues.

WORKSHEET

NAME: _____

COURSE: _____

DATE: _____

CHAPTER
10

REVIEW QUESTIONS

Based on the first part of this chapter, answer True or False to the following items:

Fertility rates are highest in Africa and the Middle East.	T	F
When nations have high fertility rates, they tend to have low infant mortality rates.	T	F
Nearly one-half of all married couples in the United States have one or two children.	T	F
In the United States, dual-career couples tend to have the fewest children.	T	F
Those who have more education tend to have fewer children.	T	F
African Americans who attended college have the same number of children as whites who attended college.	T	F
Catholics are more likely than Protestants to have three or more children.	T	F
Men who have children before age 20 tend to have lower levels of education.	T	F

EXPLORIT QUESTIONS

1. In the preliminary part of this chapter, the GSS file was used to examine the number of children that people think is ideal for a family to have. Let's use the GLOBAL file to see how the United States compares to other nations on this issue. The variable LARGE FAML measures the percent of people who think the ideal number of children is three or more.

> ➤ *Data File:* **GLOBAL**
> ➤ *Task:* **Mapping**
> ➤ *Variable 1:* **6) LARGE FAML**
> ➤ *View:* **List: Rank**

Chapter 10: Fertility and Family Life

a. People in what three countries are *most* likely to think the ideal number of children is three or more?

	NATION	PERCENT
1.	_____	_____
2.	_____	_____
3.	_____	_____

b. People in what four countries are *least* likely to think the ideal number of children is three or more?

	NATION	PERCENT
38.	_____	_____
39.	_____	_____
40.	_____	_____
41.	_____	_____

2. Earlier in this chapter we found that fertility rates were inversely (negatively) related to gross national product per capita. Do people in poorer nations simply desire to have more children than people in wealthier nations, or is there more to it than this?

Data File:	**GLOBAL**
➤ *Task:*	**Scatterplot**
➤ *Dependent Variable:*	**6) LARGE FAML**
➤ *Independent Variable:*	**22) $ PER CAP**
➤ *Display:*	**Reg. Line**

a. What is Pearson's correlation coefficient? $r =$ _____

b. Is the correlation coefficient statistically significant? Yes   No

c. These results support the claim that people in poor countries desire to have more children than people in wealthier countries. T   F

3. If the average person in a country thinks that fewer children is better than more children, will contraceptive use in that country be higher? Similarly, will abortion rates be higher?

Data File:	**GLOBAL**
➤ *Task:*	**Correlation**
➤ *Variables:*	**6) LARGE FAML**
	9) CONTRACEPT
	10) ABORTION

a. What is the correlation coefficient between LARGE FAML and CONTRACEPT?   r = _____

   Is this correlation coefficient statistically significant?   Yes   No

b. What is the correlation coefficient between LARGE FAML and ABORTION?   r = _____

   Is this correlation coefficient statistically significant?   Yes   No

c. Countries having higher percentages of people who think that the ideal number of children is three or more tend to have lower contraceptive usage.   T   F

d. Countries having higher percentages of people who think that the ideal number of children is three or more tend to have higher abortion rates.   T   F

e. Based on these results, one might conclude that abortion is used in many countries as a means to keep family sizes at a desirable level.   T   F

4. Let's take a look at what Americans think about the issue of abortion. The GSS asks respondents whether they believe abortion should be legal in a number of different circumstances. Use the UNIVARIATE task to examine each of the variables below and fill in the percentage who believe that abortion should be legal for each circumstance.

VARIABLE NAME AND DESCRIPTION   **% YES**

106) ABORT.DEF.: If there is a strong chance of serious defect in the body?   _____

107) ABORT.HLTH: If the woman's own health is seriously endangered?   _____

108) ABORT RAPE: If she became pregnant as a result of the rape?   _____

109) ABORT NO$: If the family has a very low income and cannot afford more children?   _____

110) ABORT SING: If she is not married and does not want to marry the man?   _____

111) ABORT ANY: If the woman wants it for any reason?   _____

a. For which variables did a majority of respondents believe abortion should be legal?

b. For which variables did less than a majority believe abortion should be legal?

c. How would you explain these results?

5. Now, let's use the auto-analyzer to look for variations in abortion attitudes. Return to the menu and select the AUTO-ANALYZER task. Enter 111) ABORT ANY for the variable—we will use this as an indicator of pro-choice attitudes. Select the appropriate view to answer each of the following questions.

a. In which marital category are there the most people who believe abortion should be legal for any reason? (circle one)
    1. Married
    2. Divorced/Separated
    3. Never Married

b. Those under age 30 are the most likely to be pro-choice.   T   F

c. Those who went to college (circle one)
    1. are more likely to believe abortion should be legal for any reason.
    2. are less likely to believe abortion should be legal for any reason.
    3. are no more or less likely to believe abortion should be legal for any reason.

d. Statistically speaking, women are more likely than men to be pro-choice.   T   F

e. With regard to political party affiliation, _____ are the most likely to be pro-choice. (circle one)
    1. Democrats
    2. Independents
    3. Republicans

Marriage and Family

f. With regard to religion, _____ are the most likely to be pro-choice. (circle one)
  1. liberal Protestants
  2. those who are Jewish
  3. those who have no religious affiliation

g. With regard to religion, _____ are the least likely to be pro-choice. (circle one)
  1. liberal Protestants
  2. conservative Protestants
  3. Catholics

h. Which of the above results were different from what you expected? Explain your answer.

6. The GSS data file has some lifestyle questions that allow us to examine the impact children have on people's lives. For example, how does the number of children someone has affect whether or not he or she has attended a concert in the previous year? Those over age 50 will be excluded from the analysis since they are likely to have grown children who no longer live at home.

> Data File: **GSS**
> ➤ Task: **Cross-tabulation**
> ➤ Row Variable: **84) POP MUSIC**
> ➤ Column Variable: **13) # CHILDREN**
> ➤ Subset Variable: **24) AGE**
> ➤ Subset Category: **Exclude: 3) 50 and Up**
> ➤ View: **Tables**
> ➤ Display: **Column %**

a. Fill in the percentaged results for the *top* row of this table.

	NONE	1–2	3 AND UP
YES	_____%	_____%	_____%

b. What is the value of V?   V = _____

   c. Is V statistically significant?   Yes   No

   d. Those who have more children are (circle one)

        1.   more likely to go to a concert.

        2.   less likely to go to a concert.

        3.   no more or less likely to go to a concert.

   e. With regard to concert attendance, the greatest difference is between those who have no children and those who have at least one.   T   F

7.   Are those who have more children less likely to have gone out to see a movie in the previous week?

>   *Data File:* **GSS**
>   *Task:* **Cross-tabulation**
>   ➤ *Row Variable:* **86) SEE FILM**
> ➤ *Column Variable:* **13) # CHILDREN**
> ➤ *Subset Variable:* **24) AGE**
> ➤ *Subset Category:* **Exclude: 3) 50 and Up**
>   ➤ *View:* **Tables**
>   ➤ *Display:* **Column %**

   a. Fill in the percentaged results for the *top* row of this table.

	NONE	1–2	3 AND UP
YES	_____%	_____%	_____%

   b. What is the value of V?   V = _____

   c. Is V statistically significant?   Yes   No

   d. Those who have more children are (circle one)

        1.   more likely to go to a movie.

        2.   less likely to go to a movie.

        3.   no more or less likely to go to a movie.

   e. With regard to going out to a movie, the greatest difference is between those who have no children and those who have at least one.   T   F

8. Are those who have more children less likely to have eaten out in the previous week?

      Data File: **GSS**
           Task: **Cross-tabulation**
      ➤ *Row Variable:* **85) EAT OUT**
      ➤ *Column Variable:* **13) # CHILDREN**
      ➤ *Subset Variable:* **24) AGE**
      ➤ *Subset Category:* **Exclude: 3) 50 and Up**
           ➤ *View:* **Tables**
           ➤ *Display:* **Column %**

    a. Fill in the percentaged results for the *top* row of this table.

	NONE	1–2	3 AND UP
YES	_____%	_____%	_____%

    b. What is the value of V?   V = _____

    c. Is V statistically significant?   Yes   No

    d. Those who have children are (circle one)
        1. more likely to eat out.
        2. less likely to eat out.
        3. no more or less likely to eat out.

9. Are those who have more children less likely to visit regularly with friends?

      Data File: **GSS**
           Task: **Cross-tabulation**
      ➤ *Row Variable:* **80) SOC.FRIEND**
      ➤ *Column Variable:* **13) # CHILDREN**
      ➤ *Subset Variable:* **24) AGE**
      ➤ *Subset Category:* **Exclude: 3) 50 and Up**
           ➤ *View:* **Tables**
           ➤ *Display:* **Column %**

    a. Fill in the percentaged results for the *top* row of this table.

	NONE	1–2	3 AND UP
DAILY/WKLY	_____%	_____%	_____%

    b. What is the value of V?   V = _____

    c. Is V statistically significant?   Yes   No

d. Those who have children are (circle one of the following)

1. more likely to socialize at least weekly with friends.

2. less likely to socialize at least weekly with friends.

3. no more or less likely to socialize at least weekly with friends.

e. With regard to socializing with friends, the greatest difference is between those who have no children and those who have at least one. T F

10. Finally, let's look at the relationship between happiness and having children.

Data File: **GSS**
Task: **Cross-tabulation**
➤ Row Variable: **73) HAPPY?**
➤ Column Variable: **13) # CHILDREN**
➤ Subset Variable: **24) AGE**
➤ Subset Category: **Exclude: 3) 50 and Up**
➤ View: **Tables**
➤ Display: **Column %**

a. Fill in the percentaged results for the *top* row of this table.

	NONE	1–2	3 AND UP
VERY HAPPY	_____%	_____%	_____%

b. What is the value of Cramer's V? V = _____

c. There is not a strong relationship between # CHILDREN and HAPPY?. T F

11. Based on the results for Questions 6–10, summarize the effects children have on lifestyles and personal happiness.

12. How many children, if any, would you like to have? Explain the basis for your answer.

CHAPTER 11

CHILDREARING AND SOCIALIZATION

Tasks: Cross-tabulation, Univariate
Data Files: GSS, CULTURES

Advice on raising children seems to be everywhere. Books and newspaper columns contain the latest theories on raising happy, healthy, well-adjusted children. Radio and television talk shows encourage people to call in and talk to an expert about childrearing issues. Movies and television shows often end with an insight on how to raise children. And, of course, friends, neighbors, and family are all eager to share the wisdom gained from their own experiences. The amount of information available on childrearing tells us two things—that raising children is a challenge in our society and that parents are very concerned about raising their children well. In this chapter, we will explore some of the ways that parents influence their children, maybe without even knowing it. And we will examine some of the social forces that influence our parenting styles.

Let's begin by looking at how parents influence their children through the behaviors and values they model every day. For example, most parents would like for their children to graduate from high school and, hopefully, go to college—but not all children do. In the GSS we have seen that some respondents quit school before graduating from high school and that others graduated from high school but never went to college. So, what accounts for this difference in educational outcomes? Why did some go to college while others did not? Part of the answer may lie in the example set by their parents.

> ➤ *Data File:* **GSS**
> ➤ *Task:* **Cross-tabulation**
> ➤ *Row Variable:* **27) EDUCATION**
> ➤ *Column Variable:* **42) DAD EDUC.**
> ➤ *View:* **Tables**
> ➤ *Display:* **Column %**

EDUCATION   by   DAD EDUC.
Cramer's V. 0.181 **

		DAD EDUC.				
		NO HS GRAD	HS GRAD	COLL EDUC.	Missing	TOTAL
EDUCATION	NO HS GRAD	351	84	289	252	724
		23.0%	6.8%	12.9%		14.5%
	HS GRAD	541	449	462	222	1452
		35.5%	36.5%	20.7%		29.1%
	COLL EDUC	633	698	1485	162	2816
		41.5%	56.7%	66.4%		56.4%
	Missing	3	3	10	5	21
	TOTAL	1525	1231	2236	641	4992
		100.0%	100.0%	100.0%		

Clearly there is a relationship between a respondent's level of education and the level of education attained by his or her father. The highest percentage of respondents who did not complete high school is for those whose fathers did not complete high school (23 percent). Conversely, those whose fathers went to college were far more likely to go to college (66.4 percent) than those whose fathers graduated from high school but did not go to college (56.7 percent) or who dropped out of high school (41.5 percent). Although nearly half of those whose fathers did not go to college were still able to go to college themselves, having a father who attended college does make it more likely that the respondent will have furthered his or her education after high school. The relationship between these two variables is statistically significant (V = 0.181**).

Let's see if the same relationship exists between respondent's education and mother's level of education.

Data File: **GSS**
Task: **Cross-tabulation**
Row Variable: **27) EDUCATION**
➤ Column Variable: **43) MOM EDUC.**
➤ View: **Tables**
➤ Display: **Column %**

EDUCATION by MOM EDUC.
Cramer's V: 0.241 **

		MOM EDUC.				
		NO HS GRAD	HS GRAD	COLL EDUC	Missing	TOTAL
E D U C A T I O N	NO HS GRAD	420	138	155	263	713
		26.5%	7.4%	9.9%		14.2%
	HS GRAD	575	653	251	195	1479
		36.3%	35.0%	16.1%		29.5%
	COLL EDUC	587	1074	1156	161	2817
		37.1%	57.6%	74.0%		56.2%
	Missing	6	6	2	7	21
	TOTAL	1582	1865	1562	626	5009
		100.0%	100.0%	100.0%		

The relationship between the respondent's level of education and his or her mother's level of education is even stronger than the results for the father's level of education (V = 0.241**). A mother's level of education is a good predictor of her children's eventual level of education.

Is the relationship between respondent's education and mother's education the same for males as it is for females? Let's find out.

Data File: **GSS**
Task: **Cross-tabulation**
Row Variable: **27) EDUCATION**
Column Variable: **43) MOM EDUC.**
➤ Control Variable: **25) SEX**
➤ View: **Tables (MALE)**
➤ Display: **Column %**

EDUCATION by MOM EDUC.
Controls: SEX: MALE
Cramer's V: 0.236 **

		MOM EDUC.				
		NO HS GRAD	HS GRAD	COLL EDUC	Missing	TOTAL
E D U C A T I O N	NO HS GRAD	172	72	69	114	313
		27.7%	8.4%	9.8%		14.4%
	HS GRAD	214	276	108	83	598
		34.4%	32.4%	15.3%		27.4%
	COLL EDUC	236	505	529	74	1270
		37.9%	59.2%	74.9%		58.2%
	Missing	1	2	2	4	9
	TOTAL	622	853	706	275	2181
		100.0%	100.0%	100.0%		

The results for males are almost identical to the results in the previous table. Among those whose mothers went to college, 74.9 percent went to college themselves, compared to 59.2 percent for those whose mothers just graduated high school and 37.9 percent for those whose mothers did not complete high school. The relationship is statistically significant (V = 0.236**).

Data File: **GSS**
Task: **Cross-tabulation**
Row Variable: **27) EDUCATION**
Column Variable: **43) MOM EDUC.**
Control Variable: **25) SEX**
➤ View: **Tables (FEMALE)**
➤ Display: **Column %**

		MOM EDUC.				
		NO HS GRAD	HS GRAD	COLL EDUC	Missing	TOTAL
EDUCATION	NO HS GRAD	248	66	86	149	400
		25.8%	6.5%	10.0%		14.1%
	HS GRAD	361	377	143	112	881
		37.6%	37.3%	16.7%		31.2%
	COLL EDUC	351	569	627	87	1547
		36.6%	56.2%	73.2%		54.7%
	Missing	5	4	0	3	12
	TOTAL	960	1012	856	351	2828
		100.0%	100.0%	100.0%		

The results are virtually the same for females as they were for males (V = 0.245**). Both females and males are strongly influenced by their parents' educational experiences when it comes to their own educational attainment. Having parents who went to college does not guarantee that someone will go to college—but it certainly makes it more likely.

Of course, the reason that respondents whose parents went to college are more likely to go to college themselves may be more economic than it is modeling. After all, parents who went to college are likely to have better-paying jobs and thus be better able to afford to send their children to college. So, let's turn our attention to something that is less directly tied to social class—like religion. As an adult, respondents are free to decide for themselves how often they want to go to church and what type of church, if any, they want to attend. How are people's religious choices influenced by their socialization as a child?

We'll start with religious affiliation.

Data File: **GSS**
Task: **Cross-tabulation**
➤ Row Variable: **48) RELIGION**
➤ Column Variable: **52) MOMS RELIG**
➤ View: **Tables**
➤ Display: **Column %**

		MOMS RELIG				
		PROTESTANT	CATHOLIC	JEWISH	Missing	TOTAL
RELIGION	LIB. PROT.	229	19	2	898	250
		36.9%	5.1%	10.0%		24.6%
	CON.PROT.	302	25	0	1297	327
		48.6%	6.6%	0.0%		32.2%
	CATHOLIC	20	279	0	1085	299
		3.2%	74.2%	0.0%		29.4%
	JEWISH	2	1	16	94	19
		0.3%	0.3%	80.0%		1.9%
	NONE	68	52	2	672	122
		11.0%	13.8%	10.0%		12.0%
	Missing	83	16	2	485	586
	TOTAL	621	376	20	4531	1017
		100.0%	100.0%	100.0%		

Remember to remove the control variable before continuing.

Before making our comparisons, you may notice that the variable for the respondent's own religious affiliation has two categories of Protestants, liberal and conservative, whereas there is only one Protestant category for 52) MOMS RELIG. Collapsing the two Protestant categories into one category will make our comparison more precise.

Data File: **GSS**
Task: **Cross-tabulation**
Row Variable: **48) RELIGION**
Column Variable: **52) MOMS RELIG**
View: **Tables**
Display: **Column %**

RELIGION by MOMS RELIG
Cramer's V: 0.802 **

		MOMS RELIG				
		PROTESTANT	CATHOLIC	JEWISH	Missing	TOTAL
RELIGION	PROTESTANT	531	44	2	2195	577
		85.5%	11.7%	10.0%		56.7%
	CATHOLIC	20	279	0	1085	299
		3.2%	74.2%	0.0%		29.4%
	JEWISH	2	1	16	94	19
		0.3%	0.3%	80.0%		1.9%
	NONE	68	52	2	672	122
		11.0%	13.8%	10.0%		12.0%
	Missing	83	16	2	485	586
	TOTAL	621	376	20	4531	1017
		100.0%	100.0%	100.0%		

To collapse the two categories, click on the categories labeled LIB. PROT. and CON.PROT. When these rows are highlighted, click the [Collapse] button, type in a new category label PROTESTANT, and click [OK].

Now it is easy to see the relationship between a respondent's religious affiliation and that of his or her mother. Among those whose mothers were Protestant, 85.5 percent identify themselves as Protestant. Similarly, 74.2 percent of those whose mothers were Catholic are Catholic themselves, and 80 percent of those whose mothers were Jewish say they also are Jewish. People overwhelmingly identify with their mother's religion (V = 0.802**).

In Chapter 7 we saw that the religious preferences for a respondent's mother and father were strongly correlated, so we can assume that the results for the father's religious preference will be very similar to the results in the previous table. Let's move on and look at actual church attendance. After all, it is easier to identify with a particular religion than it is to actually attend religious services.

Data File: **GSS**
Task: **Cross-tabulation**
➤ Row Variable: **50) ATTEND**
➤ Column Variable: **55) MA ATTEND**
➤ View: **Tables**
➤ Display: **Column %**

ATTEND by MA ATTEND
Cramer's V: 0.158 **

		MA ATTEND				
		RARELY	INFREQUENT	OFTEN	Missing	TOTAL
ATTEND	NEVER	62	38	99	930	199
		31.8%	14.8%	15.3%		18.1%
	MONTH/YRLY	94	151	290	2152	535
		48.2%	58.8%	44.8%		48.6%
	WEEKLY	39	68	259	1343	366
		20.0%	26.5%	40.0%		33.3%
	Missing	1	4	10	109	124
	TOTAL	195	257	648	4534	1100
		100.0%	100.0%	100.0%		

The relationship between the respondent's and mother's church attendance is not as strong as the relationship for religious affiliation, but it is statistically significant (V = 0.158**). Those whose mothers attended church often are the most likely to attend church weekly themselves (40 percent), and those whose mothers rarely went to church are the least likely to attend church weekly (20 percent). Parents do influence the religious values and practices of their children even as adults.

So far, we have focused on how parents influence their children indirectly through modeling different behaviors and values. But parents also have values that they try to pass on directly to their children through teaching. In 1956, Melvin L. Kohn conducted a study of the values guiding childrearing practices among American parents. His findings indicated two distinctive clusters of childrearing values and showed that parents of higher social status put more stress on one while lower-status parents put more stress on the other.

Lower-status parents tended to put greater emphasis on values such as good manners, obedience, neatness, and cleanliness than did higher-status parents. In contrast, higher-status parents placed greater value on curiosity, responsibility, and consideration for others (Kohn, 1959). Kohn identified the first value cluster as concern about *conforming to norms* and the second as concern about *self-expression and self-direction*. Over the years Kohn, with various associates (especially Carmi Schooler), has restudied childrearing values and has had similar findings, even in other countries (Pearlin and Kohn, 1966; Kohn and Schooler, 1969, 1983). To explain these results, Kohn and his associates theorize as follows:

Parents try to give their children the best possible chances in life. To do so, they try to instill in them what they have learned from their life experiences about how the world works and how best to get along. Lower-status parents tend to hold the kinds of jobs in which they do well to the extent that they are regarded as dependable—as people who are prompt, who show up at work looking neat and clean, who have adequate manners, and who obey the rules. In contrast, higher-status parents have found that they are more successful in their jobs when they can take individual initiative, are responsible and curious, and have good interpersonal skills. So parents draw on their own experiences in deciding how to raise their children. To the extent that children grow up and take jobs similar to those held by their parents, such a pattern may be quite functional. On the other hand, it might cause people simply to end up in the same kinds of jobs as their parents because of the ways they were socialized.

Keep in mind that these patterns are simply tendencies—*all* parents probably value all of the qualities presented to them in Kohn's research. But as we take a look at some similar childrearing values from the GSS data set, we will expect to find the same patterns along social status lines.

To begin, let's look at two measures which assess the importance of different values that children should be taught. The first question asks whether obedience is the most important thing that parents should teach their children. The second question asks the same thing about teaching children to think for themselves. Kohn's theory would predict that **lower-status parents will believe parents should stress obedience, while upper-status parents will believe that parents should stress independent thinking.**

	Data File:	**GSS**
	Task:	**Cross-tabulation**
➤	Row Variable:	**113) KID OBEY**
➤	Column Variable:	**28) INCOME**
➤	View:	**Tables**
➤	Display:	**Column %**

KID OBEY   by   INCOME
Cramer's V: 0.127 **

		INCOME				
		LOW	MIDDLE	HIGH	Missing	TOTAL
KID OBEY	MOST IMP.	235	241	147	121	623
		26.3%	18.7%	13.4%		19.0%
	LESS IMP.	660	1050	946	343	2656
		73.7%	81.3%	86.6%		81.0%
	Missing	466	698	516	226	1906
	TOTAL	895	1291	1093	690	3279
		100.0%	100.0%	100.0%		

As you might expect, there is a clear difference along class lines. As social status increases, the emphasis placed on obedience declines. Parents from the lowest income category (26.3 percent) are twice as likely as parents from the highest income category (13.4 percent) to believe that obedience is the most important thing to teach a child. The difference is statistically significant (V = 0.127**). Now let's test the second half of our hypothesis.

Data File: **GSS**
Task: **Cross-tabulation**
➤ Row Variable: **114) THINK SELF**
➤ Column Variable: **28) INCOME**
➤ View: **Tables**
➤ Display: **Column %**

THINK SELF   by   INCOME
Cramer's V: 0.136 **

		INCOME				
		LOW	MIDDLE	HIGH	Missing	TOTAL
THINK SELF	MOST IMP.	361	650	633	187	1644
		40.3%	50.3%	57.9%		50.1%
	LESS IMP.	534	641	460	277	1635
		59.7%	49.7%	42.1%		49.9%
	Missing	466	698	516	226	1906
	TOTAL	895	1291	1093	690	3279
		100.0%	100.0%	100.0%		

Again, there is a statistically significant difference that supports the hypothesis (V = 0.136**). Parents in the highest income category (57.9 percent) are the most likely, and those in the lowest income category are the least likely (40.3 percent), to believe that the most important thing you can teach children is to think for themselves. Thus, our hypothesis is supported.

Because education is highly correlated with social class, and because colleges tend to emphasize independent thinking, we would expect similar results using education as the independent variable.

Data File: **GSS**
Task: **Cross-tabulation**
Row Variable: **114) THINK SELF**
➤ Column Variable: **27) EDUCATION**
➤ View: **Tables**
➤ Display: **Column %**

THINK SELF   by   EDUCATION
Cramer's V: 0.229 **

		EDUCATION				
		NO HS GRAD	HS GRAD	COLL EDUC	Missing	TOTAL
THINK SELF	MOST IMP.	178	508	1142	3	1828
		27.9%	44.5%	58.6%		49.0%
	LESS IMP.	461	633	808	10	1902
		72.1%	55.5%	41.4%		51.0%
	Missing	337	533	1028	8	1906
	TOTAL	639	1141	1950	21	3730
		100.0%	100.0%	100.0%		

The results here are even more pronounced than those in the previous table. Those who did not graduate from high school (27.9 percent) are far less likely than those who did graduate from high school (44.5 percent) or those who went to college (58.6 percent) to believe that parents should stress independent thinking. It would appear that educational experiences are at least as important as, if not more important than, social class in shaping our childrearing practices.

Let's shift our attention to childrearing practices in preindustrial societies. An advantage of using anthropological data, such as that provided in the CULTURES file, is that it's based on actual observations. Respondents in surveys, on the other hand, sometimes provide answers that are socially acceptable rather than giving responses that truly reflect their opinions or behaviors. Anthropological data further allow us to examine societies prior to the influence of industrialization.

More than 100 societies are coded in regard to the interpersonal warmth parents express to their children. They are also coded in terms of whether competitiveness among children was emphasized in the society. Are societies in which parents are warm to their children more or less likely to emphasize competitiveness among boys?

➤ *Data File:* **CULTURES**
➤ *Task:* **Cross-tabulation**
➤ *Row Variable:* **38) COMPETE B.**
➤ *Column Variable:* **32) WARM:KIDS**
➤ *View:* **Tables**
➤ *Display:* **Column %**

COMPETE B.   by   WARM:KIDS
Cramer's V: 0.325 *
Warning: Potential significance problem. Check row and column totals.

		COOL	WARMER	VERY WARM	Missing	TOTAL
COMPETE B.	LOW	3	4	21	20	28
		16.7%	30.8%	44.7%		35.9%
	MEDIUM	3	4	15	20	22
		16.7%	30.8%	31.9%		28.2%
	HIGHER	6	2	10	12	18
		33.3%	15.4%	21.3%		23.1%
	VERY HIGH	6	3	1	6	10
		33.3%	23.1%	2.1%		12.8%
	Missing	7	3	10	30	50
	TOTAL	18	13	47	88	78
		100.0%	100.0%	100.0%		

(WARM:KIDS spans COOL, WARMER, VERY WARM columns)

Focus on the bottom row of the table. Societies in which parents are cool to their children (33.3 percent) are much more likely than societies in which parents are very warm to their children (2.1 percent) to place a very high emphasis on competitiveness. The top row of the table presents a parallel picture. Societies in which parents are very warm to their children are much more likely to place a low emphasis on competitiveness. The value for Cramer's V is fairly strong and the results are statistically significant (V = 0.325*).

Notice that the screen is warning us that there may be a significance problem, and to "check the row and column totals." In simple terms, this means that there are too many possible categories for the number of cases being included in the analysis—hence, the probability value might not be accurate. This usually isn't a major concern, but one way to deal with it is to collapse, or combine, the categories for one or both variables used in the analysis. To do this, click on the category labels HIGHER and VERY HIGH. When these rows are highlighted, click the [Collapse] button, type in a new category label HIGH, and click [OK]. Notice that these two categories are now combined in the table. You can follow this same procedure for combining the first two categories too. Click the labels LOW and MEDIUM, select [Collapse], use LOW as the label for the combined categories, and click [OK]. Now you should see the following table:

Data File: **CULTURES**
Task: **Cross-tabulation**
Row Variable: **38) COMPETE B.**
Column Variable: **32) WARM:KIDS**
View: **Tables**
Display: **Column %**

COMPETE B.   by   WARM:KIDS
Cramer's V: 0.369 **

		COOL	WARMER	VERY WARM	Missing	TOTAL
COMPETE B.	LOW	6	8	36	40	50
		33.3%	61.5%	76.6%		64.1%
	HIGH	12	5	11	18	28
		66.7%	38.5%	23.4%		35.9%
	Missing	7	3	10	30	50
	TOTAL	18	13	47	88	78
		100.0%	100.0%	100.0%		

(WARM:KIDS spans COOL, WARMER, VERY WARM columns)

Notice that the relationship in this modified table remains strong and statistically significant (V = 0.369**). Also notice that the warning message about the significance problem has disappeared.

Does a similar relationship exist between parental warmth and competitiveness among girls?

Data File: **CULTURES**
Task: **Cross-tabulation**
➤ Row Variable: **39) COMPETE G.**
➤ Column Variable: **32) WARM:KIDS**
➤ View: **Tables**
➤ Display: **Column %**

COMPETE G.   by   WARM:KIDS
Cramer's V: 0.327 *
Warning: Potential significance problem. Check row and column totals.

		COOL	WARMER	VERY WARM	Missing	TOTAL
COMPETE G.	LOW	3	3	24	19	30
		18.8%	25.0%	54.5%		41.7%
	MEDIUM	4	5	12	23	21
		25.0%	41.7%	27.3%		29.2%
	HIGHER	6	2	8	9	16
		37.5%	16.7%	18.2%		22.2%
	VERY HIGH	3	2	0	4	5
		18.8%	16.7%	0.0%		6.9%
	Missing	9	4	13	33	59
	TOTAL	16	12	44	88	72
		100.0%	100.0%	100.0%		

Examine the top and bottom rows and you'll see that a similar relationship does indeed exist. (If you like, collapse the categories for COMPETE G. as done in the previous example.) Overall, it appears that societies in which parents are very warm to their children place less emphasis on competitiveness.

In the United States there has been much discussion in recent decades about whether corporal punishment (e.g., spanking) should be used as a means of disciplining a child. There isn't a variable in the CULTURES file that specifically measures spanking, but there is one that measures how often parents are hostile and aggressive toward their children.

Parents in about half the societies (53.1 percent) are rarely hostile toward their children; parents often direct hostility toward their children in 11.5 percent of societies. Let's see if societies in which parents neglect their children are the same societies in which parents are often hostile toward their children.

Data File: **CULTURES**
Task: **Cross-tabulation**
➤ Row Variable: **33) HIT KIDS**
➤ Column Variable: **35) C.NEGLECT**
➤ View: **Tables**
➤ Display: **Column %**

HIT KIDS   by   C.NEGLECT
Cramer's V: 0.313 **
Warning: Potential significance problem. Check row and column totals.

		RARELY	SOMETIMES	OFTEN	Missing	TOTAL
HIT KIDS	RARELY	31	9	1	19	41
		53.4%	42.9%	25.0%		49.4%
	SOMETIMES	23	7	0	10	30
		39.7%	33.3%	0.0%		36.1%
	OFTEN	4	5	3	1	12
		6.9%	23.8%	75.0%		14.5%
	Missing	10	2	3	58	73
	TOTAL	58	21	4	88	83
		100.0%	100.0%	100.0%		

Examine the third column of the table and you'll see that societies in which parents often neglect their children (75 percent) are the same societies in which parents are often hostile and aggressive toward their children. In societies where parents rarely neglect their children, only 6.9 percent often direct hostility toward children. Neglect and hostility seem to go hand in hand.

When you were growing up, do you think the school bully or the toughest, most aggressive athlete in school had parents who treated him warmly, or do you think he was treated coolly by his parents? We can't test this exact concept, but we can use this data set to assess the general relationship between parental warmth and the emphasis that parents put on aggressiveness in boys.

Data File: **CULTURES**
Task: **Cross-tabulation**
➤ Row Variable: **43) TUFF BOYS***
➤ Column Variable: **32) WARM:KIDS**
➤ View: **Tables**
➤ Display: **Column %**

TUFF BOYS*   by   WARM:KIDS
Cramer's V: 0.351 *

		WARM:KIDS				
		COOL	WARMER	VERY WARM	Missing	TOTAL
TUFF BOYS*	LOW	3	6	23	23	32
		18.8%	54.5%	60.5%		49.2%
	HIGH	13	5	15	20	33
		81.3%	45.5%	39.5%		50.8%
	Missing	9	5	19	45	78
	TOTAL	16	11	38	88	65
		100.0%	100.0%	100.0%		

Again, if you examine the second row of the table, you'll see a trend. Societies in which parents are very warm to their children (39.5 percent) are less likely than societies in which parents are cool to their children (81.3 percent) to emphasize aggressiveness and overall "toughness" among boys. If you repeat this analysis using 45) TUFF GIRL* as the row variable, you will find a similar relationship among girls.

In the worksheet section that follows, you will examine several other issues related to childrearing using the General Social Survey.

References

Kohn, Melvin. 1959. "Social Class and Parental Values," *American Journal of Sociology*, 64:337–351.

Kohn, Melvin, and Carmi Schooler. 1969. "Class, Occupation, and Orientation," *American Sociological Review*, 34:659–678.

———. 1983. *Work and Personality*. Norwood, NJ: Ablex.

Pearlin, L. I., and Melvin Kohn. 1966. "Social Class, Occupation, and Parental Values: A Cross-National Study," *American Sociological Review*, 31:466–479.

WORKSHEET

NAME: _____

COURSE: _____

DATE: _____

Workbook exercises and software are copyrighted. Copying is prohibited by law.

REVIEW QUESTIONS

Based on the first part of this chapter, answer True or False to the following items:

The educational attainment of males is related to the educational attainment of their mothers.	T	F
The majority of Americans are affiliated with the same religion as their parents.	T	F
There is not a statistically significant relationship between someone's own frequency of church attendance and that of his or her parents.	T	F
Melvin Kohn believed that higher-status parents are primarily concerned with teaching their children to be obedient.	T	F
Those who have higher incomes are the most likely to emphasize the importance of obedience.	T	F
People with more education tend to emphasize the importance of thinking for oneself.	T	F
Societies in which parents are warm to their children are less likely to place an emphasis on competitiveness and aggressiveness.	T	F
Societies in which parents neglect their children are the same societies in which parents are often hostile and aggressive toward their children.	T	F

EXPLORIT QUESTIONS

1. What do Americans believe about spanking as a form of discipline? Let's return to the GSS data file and examine this issue.

> *Data File:* **GSS**
> *Task:* **Univariate**
> *Primary Variable:* **112) SPANKING**
> *View:* **Pie**

a. What percentage of U.S. respondents agree that sometimes it is necessary to spank a child? _____%

b. How would you answer this question?　　　　　　　　Agree　　Disagree

2. Is spanking an old-fashioned practice that is no longer approved of by young people? To find out, test the hypothesis that **older people will be more likely than those who are younger to approve of spanking as a form of discipline**.

> *Data File:* **GSS**
> ➤ *Task:* **Cross-tabulation**
> ➤ *Row Variable:* **112) SPANKING**
> ➤ *Column Variable:* **24) AGE**
> ➤ *View:* **Tables**
> ➤ *Display:* **Column %**

a. Fill in the percentaged results for the *top* row of the table.

	<30	30–49	50 AND UP
AGREE	_____%	_____%	_____%

b. What is the value of Cramer's V?   V = _____

c. Is V statistically significant?   Yes   No

d. Is the hypothesis supported?   Yes   No

3. Let's look at the relationship between education and approval of spanking.

> *Data File:* **GSS**
> *Task:* **Cross-tabulation**
> *Row Variable:* **112) SPANKING**
> ➤ *Column Variable:* **27) EDUCATION**
> ➤ *View:* **Tables**
> ➤ *Display:* **Column %**

a. Fill in the percentaged results for the *top* row of the table.

	NO HS GRAD	HS GRAD	COLL EDUC
AGREE	_____%	_____%	_____%

b. Is V statistically significant?   Yes   No

c. College graduates are less likely to approve of spanking.   T   F

d. How might you explain the relationship between education and approval of spanking?

4. The hypothesis is: **Men will be more likely than women to approve of spanking as a form of discipline.**

> Data File: **GSS**
> Task: **Cross-tabulation**
> Row Variable: **112) SPANKING**
> ➤ Column Variable: **25) SEX**
> ➤ View: **Tables**
> ➤ Display: **Column %**

a. What percentage of men approve of spanking?   _____%

b. What percentage of women approve of spanking?   _____%

c. Is V statistically significant?   Yes   No

d. Is the hypothesis supported?   Yes   No

e. How would you explain these results?

5. Does church attendance influence people's approval or disapproval of spanking?

> Data File: **GSS**
> Task: **Cross-tabulation**
> Row Variable: **112) SPANKING**
> ➤ Column Variable: **50) ATTEND**
> ➤ View: **Tables**
> ➤ Display: **Column %**

a. Fill in the percentaged results for this table.

	NEVER	MONTH/YRLY	WEEKLY
AGREE	_____%	_____%	_____%
DISAGREE	_____%	_____%	_____%

b. Is V statistically significant?   Yes   No

   c. Are people who attend church more frequently more likely to approve of spanking?   Yes   No

6. For those indicating a religious preference, does the fundamentalism or liberalism of the respondent's religion make a difference in approval or disapproval of spanking?

> Data File: **GSS**
> Task: **Cross-tabulation**
> Row Variable: **112) SPANKING**
> ➤ Column Variable: **49) R.FUND/LIB**
> ➤ View: **Tables**
> ➤ Display: **Column %**

   a. Fill in the percentaged results for this table.

	FUNDAMENT.	MODERATE	LIBERAL
AGREE	_____%	_____%	_____%
DISAGREE	_____%	_____%	_____%

   b. Is V statistically significant?   Yes   No

   c. Are fundamentalists more likely to approve of spanking?   Yes   No

   d. Review the results of Questions 5 and 6. How would you explain the relationship between religion and spanking?

7. Let's see how many people in the GSS were raised as an only child and how many had brothers or sisters.

> Data File: **GSS**
> ➤ Task: **Univariate**
> ➤ Primary Variable: **41) # SIBS**
> ➤ View: **Pie**

   a. What percentage of respondents were an only child?   _____%

    b. What percentage had 1 to 3 siblings?   _____%

    c. What percentage had 4 or more siblings?   _____%

8.   Let's look at the relationship between number of siblings and education. Because only children will not have to share their parents' time and resources with siblings, let's hypothesize that **only children will be more likely to have attended college.**

> *Data File:* **GSS**
> ➤ *Task:* **Cross-tabulation**
> ➤ *Row Variable:* **27) EDUCATION**
> ➤ *Column Variable:* **41) # SIBS**
> ➤ *View:* **Tables**
> ➤ *Display:* **Column %**

    a. Fill in the percentaged results for the *bottom* row of the table.

	NONE	1–3	4 AND ABOV
COLL EDUC	_____%	_____%	_____%

    b. What is the value of V?   V = _____

    c. Is V statistically significant?   Yes   No

    d. Is the hypothesis supported?   Yes   No

    e. Those who had 1 to 3 siblings most closely resemble (circle one)
        1.   those who had no siblings.
        2.   those who had 4 or more siblings.

    f. How would you explain these results?

CHAPTER 12

WORK AND FAMILY

> *Tasks:* Univariate, Historical Trends, Mapping, Scatterplot, Cross-tabulation
> *Data Files:* TRENDS, STATES, GLOBAL, CULTURES, GSS

Most of the contemporary debate surrounding family gender roles has centered on the roles of women. In Western countries, since the early part of the 20th century, it has been generally accepted as natural that men would leave their homes to provide for their families. Women have had to balance their desire to be employed—or the economic necessity of employment—against a cultural expectation that their primary responsibility is to nurture their families.

Although the concept of going to work is taken for granted today, it is a relatively recent phenomenon when viewed in historical context. For thousands of years prior to industrialization, both men and women participated in the provision of food, shelter, and clothing for the family—and most of this production took place in the home. There was no discussion about women, or for that matter men, working outside the home—because most of the work that needed to be done was done at home.

Let's briefly turn to the CULTURES file (which contains data on preindustrial societies) to examine the percent of subsistence labor (labor for food, shelter, clothing) that was supplied by females.

> ➤ *Data File:* **CULTURES**
> ➤ *Task:* **Univariate**
> ➤ *Primary Variable:* **4) %FEM.SUBST**
> ➤ *View:* **Pie**

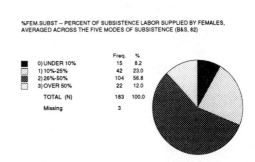

%FEM.SUBST -- PERCENT OF SUBSISTENCE LABOR SUPPLIED BY FEMALES, AVERAGED ACROSS THE FIVE MODES OF SUBSISTENCE (B&S, 82)

	Freq.	%
0) UNDER 10%	15	8.2
1) 10%-25%	42	23.0
2) 26%-50%	104	56.8
3) OVER 50%	22	12.0
TOTAL (N)	183	100.0
Missing	3	

As these results show, women contributed between 26 and 50 percent of the subsistence labor in more than half of the societies. In 12 percent of the societies, women were responsible for more than half of the subsistence labor. Keep in mind that women were also responsible for most of the child care required in these societies.

During the past 300 years, production has gradually moved from the home to economic institutions that are distinct and separate from the family. In an industrialized society, going to work generally means being away from the family. This puts families in the relatively new position of deciding who should go to work—the husband, the wife, or both? Of course, one need not be employed to be involved in economic production. Persons who are not employed may still be involved in maintaining the home

195

and producing goods and services for their families and others. But the question of how husbands and wives should divide the responsibilities for paid employment still remains.

Change to the GLOBAL file to examine modern-day nations in terms of female employment. The variable F/M EMPLOY indicates the proportion of females to males who are part of the paid labor force.

> *Data File:* **GLOBAL**
> *Task:* **Mapping**
> *Variable 1:* **73) F/M EMPLOY**
> *View:* **Map**

F/M EMPLOY -- FEMALE EMPLOYMENT AS PERCENT OF MALE EMPLOYMENT
(HDR, 1995)

The proportion of females in the paid labor force appears to be the highest in Europe, North America, East Asia, and the nations that were formerly part of the Soviet Union. Female employment is lowest in South and Central America, North Africa, and the Middle East.

F/M EMPLOY: Female employment as a percent of male employment

Data File: **GLOBAL**
Task: **Mapping**
Variable 1: **73) F/M EMPLOY**
> *View:* **List: Rank**

RANK	CASE NAME	VALUE
1	Bulgaria	88
1	Armenia	88
3	Tanzania	85
3	Benin	85
3	Mozambique	85
6	Czech Republic	84
7	Rwanda	83
7	Mongolia	83
7	Niger	83
10	Finland	82

There is substantial variation between the nations at the top of the list and those at the bottom. Bulgaria and Armenia (tied at 88 percent) are the highest in terms of the proportion of women who are employed, and numerous Muslim countries anchor the bottom of the list, including Bahrain (10 percent) and Saudi Arabia (11 percent). The United States can be found in the top third of the list with a rate of 65 percent.

The remainder of this chapter focuses on female employment in the United States. To begin, look at women's rates of participation in the paid labor force between 1900 and 2000.

> *Data File:* **TRENDS**
> > *Task:* **Historical Trends**
> *Variable:* **26) %FEM EMPL**

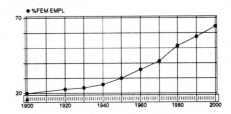

Percentage of females employed full time outside the home

As you can see in this graph, the female employment rate in the United States rose steadily during the 20th century, from 20 percent in 1900 to 64 percent in 2000. The rate of change increases notably after 1940. Many women entered the workforce during World War II, then remained employed following the war. The 1960s brought increased attention to the issue of equal rights for women in the workplace, and the upward trend in female employment continues through the 1990s.

One of the most notable changes that occurred in the past few decades is the increase in the percentage of married women who are employed outside the home. (Before continuing with the next example, delete the previously selected variable or click the [Clear All] button.)

> *Data File:* **TRENDS**
> *Task:* **Historical Trends**
> *Variable:* **27) M FEM EMP%**

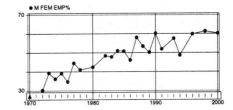

Percentage of married females employed part time or full time

This graph uses data from the General Social Survey to track changes in the labor force participation rates of married women from 1972 to 2000. As this trend shows, the percentage of married women in the paid labor force doubled in this relatively brief period of time from 30 percent in 1972 to 60 percent in 2000. For comparison, let's look at the percentage of women who categorize themselves as full-time homemakers during this same period of time.

> *Data File:* **TRENDS**
> *Task:* **Historical Trends**
> *Variable:* **28) HOMEMAKER%**

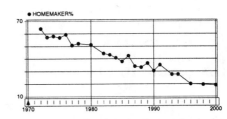

Percentage of women who say they are full-time homemakers

As you would expect given the previous graph, the percentage of women who describe themselves as full-time homemakers decreases dramatically from 1972 to 2000. In 1972, approximately 63 percent of all women said they were full-time homemakers compared to just 19 percent in 2000. In other words, women in 1972 were three times more likely than women in 2000 to identify themselves as homemakers. That is a very remarkable change for such a short period of time.

Let's examine female employment rates across the regions of the United States.

> Data File: **STATES**
> > Task: **Mapping**
> Variable 1: **41) %FEM.WRK**
> > View: **Map**

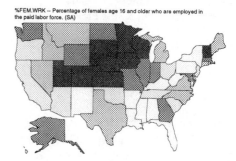

%FEM.WRK -- Percentage of females age 16 and older who are employed in the paid labor force. (SA)

This first map depicts female employment rates in 1999. You will see that female employment rates are not highly regional, except that the rates are lower in the South.

Data File: **STATES**
Task: **Mapping**
Variable 1: **41) %FEM.WRK**
> View: **List: Rank**

%FEM.WRK: Percent of females 16 and older who are employed in the paid labor force

RANK	CASE NAME	VALUE
1	Minnesota	73.0
2	Colorado	71.4
3	South Dakota	71.1
4	Nebraska	71.0
5	New Hampshire	70.3
6	Iowa	70.1
6	Wisconsin	70.1
8	Kansas	70.0
9	Vermont	69.9
10	Utah	69.6

The state with the largest percentage of employed women is Minnesota (73 percent), followed by Colorado (71.4 percent) and South Dakota (71.1 percent). Even in the state with the lowest female employment rate, West Virginia (52.7 percent), more than half of the female population is employed. As we saw in the map, 7 of the 10 states with the lowest female employment rates are in the South.

Now let's turn to the early part of the 20th century.

Data File: **STATES**
Task: **Mapping**
> Variable 1: **71) %FEM.WRK20**
> > View: **Map**

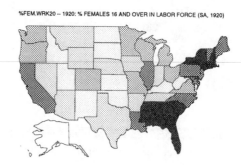

%FEM.WRK20 -- 1920: % FEMALES 16 AND OVER IN LABOR FORCE (SA, 1920)

The picture in 1920 is very different. Most of the states with the highest female employment rates are located in the Northeast and the Southeast. Look at the distribution for a more detailed analysis.

%FEM.WRK20: 1920: % Females 16 and over in labor force

Data File: **STATES**
Task: **Mapping**
Variable 1: **71) %FEM.WRK20**
➤ View: **List: Rank**

RANK	CASE NAME	VALUE
1	South Carolina	37.0
2	Rhode Island	35.7
3	Massachusetts	35.1
4	Mississippi	31.7
5	New Hampshire	30.9
6	New York	30.2
7	Connecticut	30.1
8	Georgia	29.8
9	Alabama	27.8
10	Florida	27.1

First, notice that in no state are a majority of the women employed. South Carolina's female employment rate of 37 percent was the highest of that time. This draws our attention to the second trend: Whereas in 1990 female employment is lowest in the South, in 1920 that was where many of the states with the highest rates were located. In addition to South Carolina (37.0 percent), Mississippi (31.7 percent), Georgia (29.8 percent), Alabama (27.8 percent), and Florida (27.1 percent) are all among the ten states with the highest female employment rates. The rest of the top ten states are located in the Northeast.

Part of the explanation for the shifting female employment rates in the South can be found by comparing the reasons women used to take jobs outside the home with the reasons many women are employed today. But this is best explained if we go back even further—to the 1880s. In 1880, just after the abolition of slavery, less than 15 percent of American women held jobs outside the home. This picture changed dramatically in the South over the following decades. As shown above, employment rates in southern states more than doubled by the 1920s. But the fact is, one of every five working women in the United States was African American in 1920, while only one of every ten women was African American. Or, put another way, in 1920, 44 percent of African-American women worked—not far behind the current rate today—while only 20 percent of native-born white women worked outside the home.

Let's return to our question: Why do women take jobs outside the home?

Data File: **STATES**
➤ Task: **Scatterplot**
➤ Dependent Variable: **71) %FEM.WRK20**
➤ Independent Variable: **70) F WAGES 20**
➤ View: **Reg. Line**

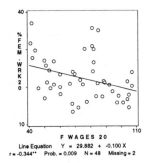

Line Equation   Y = 29.882 + -0.100 X
r = -0.344**   Prob. = 0.009   N = 48   Missing = 2

This scatterplot shows a negative correlation (r = –0.344**) between what was paid to a farm hand in 1920 and the female employment rate in 1920. In 1920, a large proportion of men who earned wages outside the home were employed as farm hands. So farm wages are a good indicator of the typical income that many men brought home from work. As we can see from the scatterplot, more women worked in states where the farm wages were lower. This would suggest that in 1920 many of the women who were employed were working only out of dire economic necessity—that is, their husbands did not make sufficient income to support the family. Let's jump ahead 20 years and conduct a similar analysis.

Data File:	**STATES**
Task:	**Scatterplot**
➤ Dependent Variable:	**68) %FEM.WRK40**
➤ Independent Variable:	**69) MEDIAN$ 40**
➤ View:	**Reg. Line**

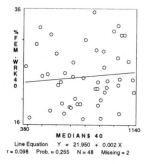

There is virtually no relationship between the female employment rate and the median income in 1940. The correlation of 0.098 is not statistically significant (Prob. = 0.255). Reasons for women's being employed other than economic necessity appear to be emerging. Now let's take a look at more recent data to see if this shift is evidence of a trend.

Data File:	**STATES**
Task:	**Scatterplot**
➤ Dependent Variable:	**41) %FEM.WRK**
➤ Independent Variable:	**38) MED.FAM$**
➤ View:	**Reg. Line**

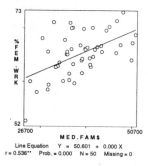

The results of this scatterplot are much more distinct and contrast starkly with the earlier findings. *There is a strong positive correlation (r = 0.536**) between the female employment rate and the median family income in 1999.* In other words, whereas high female employment rates used to be associated with economic hardship, today high female employment rates are indicative of economic opportunity. In the past, upper-income families were overwhelmingly families in which the wives were not employed. Today, they are *primarily* two-earner families.

How can we account for this change in the relationship between economic need and female employment rates? Part of the explanation is economic. In the past, men and women could secure well-paying jobs even if they did not have a great deal of formal training, but that is no longer the case. More than ever before, education and economic opportunity are closely linked. Women who are married to men with lower incomes probably have a lower earning potential themselves. Many women's earnings may not even be enough to offset the cost of child care. In contrast, women who are married to upper-income men probably have more education and better job opportunities themselves, which makes employment more attractive. Thus, the better off a woman is financially, the more likely she is to be employed.

But is this shift in women's employment patterns entirely economic, or is there an ideological component as well? In Chapter 3, we saw that gender role attitudes have changed in recent decades as an increasingly large majority of Americans approve of women being employed outside the home. Let's take a look at the GSS to see if gender role attitudes are a good predictor of the actual work patterns for married couples.

➤ *Data File:* **GSS**
➤ *Task:* **Cross-tabulation**
➤ *Row Variable:* **31) FAM.WORK**
➤ *Column Variable:* **67) WOMEN WORK**
➤ *View:* **Tables**
➤ *Display:* **Column %**

FAM.WORK   by   WOMEN WORK
Cramer's V: 0.049

		WOMEN WORK			
		APPROVE	DISAPPROVE	Missing	TOTAL
FAM.WORK	1 FULLTIME	128	29	295	157
		24.2%	29.3%		25.0%
	1 FT/1 PT	112	17	231	129
		21.2%	17.2%		20.5%
	2 FULLTIME	289	53	663	342
		54.6%	53.5%		54.5%
	Missing	981	228	2623	3832
	TOTAL	529	99	3812	628
		100.0%	100.0%		

There are a couple of interesting findings in this table. First of all, note that the total number of respondents who approve of women working outside the home (529 respondents) outnumbers those who disapprove (99 respondents) by over 5 to 1. So the vast majority of Americans do hold less traditional gender role attitudes. Secondly, look at the relationship between gender role attitudes and actual family work patterns—there is virtually no difference between those who approve and those who disapprove of women working outside the home with regard to actual employment patterns. Those who disapprove of women working outside the home (53.5 percent) are as likely as those who approve of women working outside the home (54.6 percent) to have a dual-income marriage. While ideological changes have certainly accompanied the increasing female employment rate, even those who hold more traditional gender role attitudes are likely to be in a marriage where both spouses are employed—perhaps out of economic necessity.

Let's take a further look at the family work patterns in the GSS.

Data File: **GSS**
➤ Task: **Univariate**
➤ Primary Variable: **31) FAM.WORK**
➤ View: **Pie**

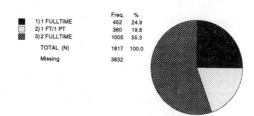

FAM.WORK -- Family work status for married respondents

	Freq.	%
■ 1) 1 FULLTIME	452	24.9
▨ 2) 1 FT/1 PT	360	19.8
▧ 3) 2 FULLTIME	1005	55.3
TOTAL (N)	1817	100.0
Missing	3832	

According to these results, most of the couples have a dual-income marriage. In 55.3 percent of the cases, both spouses are employed full-time and in 19.8 percent of the cases one spouse is employed full-time while the other is employed part-time. About one-fourth of the respondents have a single economic provider for their family.

Based on the state-level trend we saw earlier, we would expect dual-income families to have higher incomes.

Data File: **GSS**
➤ Task: **Cross-tabulation**
➤ Row Variable: **28) INCOME**
➤ Column Variable: **31) FAM.WORK**
➤ View: **Tables**
➤ Display: **Column %**

INCOME   by   FAM.WORK
Cramer's V: 0.163 **

		FAM.WORK				
		1 FULLTIME	1 FT/1 PT	2 FULLTIME	Missing	TOTAL
INCOME	LOW	48	22	23	1268	93
		11.8%	6.8%	2.5%		5.7%
	MIDDLE	185	122	288	1394	595
		45.6%	37.9%	31.4%		36.2%
	HIGH	173	178	605	653	956
		42.6%	55.3%	66.0%		58.2%
	Missing	46	38	89	517	690
	TOTAL	406	322	916	3832	1644
		100.0%	100.0%	100.0%		

As expected, families who have only one wage-earner are the least likely to be in the high income category (42.6 percent) while those who have two full-time wage-earners are the most likely to be in this category (66 percent). Families who have one full-time and one part-time wage-earner fall midway in between (55.3 percent). So, dual-income families tend to be well-off financially (V = 0.163**).

Of course, a family's employment pattern is not static. There are many circumstances that may affect the work status of one or both spouses—such as childrearing, illness, or being laid-off. Let's look at the first of these factors—childrearing. How does the stage a family is in with regard to having or raising children influence their family work status?

<table>
<tr><td>Data File: GSS</td></tr>
</table>

Data File: **GSS**
Task: **Cross-tabulation**
➤ Row Variable: **31) FAM.WORK**
➤ Column Variable: **11) FAM.STAGE**
➤ View: **Tables**
➤ Display: **Column %**

FAM.WORK   by   FAM.STAGE
Cramer's V: 0.134 **

		FAM.STAGE				
		NO KIDS	WITH KIDS	EMPTY NEST	Missing	TOTAL
FAM.WORK	1 FULLTIME	18	304	108	22	430
		8.8%	28.3%	23.3%		24.7%
	1 FT/1 PT	32	246	75	7	353
		15.6%	22.9%	16.2%		20.3%
	2 FULLTIME	155	523	280	47	958
		75.6%	48.7%	60.5%		55.0%
	Missing	40	184	534	3074	3832
	TOTAL	205	1073	463	3150	1741
		100.0%	100.0%	100.0%		

Those who are married and have children are the most likely to have a one-income family (28.3 percent) and the least likely to have a two-income family (48.7 percent). Conversely, those who are married but have no children are the least likely to have only one spouse employed (8.8 percent) and the most likely to have both spouses employed (75.6 percent). Those who are in the empty-nest stage fall midway in between for both categories. Although nearly one-half of the married couples with children still have a two-income family, it appears that many couples go from being a two-income family to a one-income family during the childrearing years. After the children leave home, many of these couples are likely to again become a two-income family.

Supporting a family financially is just one of the responsibilities that must be assumed by married couples. There also are the myriad of household chores such as cooking, cleaning, shopping, and maintenance that must be performed. Who does most of the work of maintaining the household?

Data File: **GSS**
Task: **Cross-tabulation**
➤ Row Variable: **69) DOMES.DUTY**
➤ Column Variable: **25) SEX**
➤ Subset Variable: **1) MARITAL**
➤ Subset Category: **Include: 1) Married**
➤ View: **Tables**
➤ Display: **Column %**

DOMES.DUTY   by   SEX

Cramer's V: 0.693 **

		SEX		
		MALE	FEMALE	TOTAL
DOMES.DUTY	YES,SOLELY	19	241	260
		8.1%	71.3%	45.3%
	YES,EQUALY	103	90	193
		43.6%	26.6%	33.6%
	NO	114	7	121
		48.3%	2.1%	21.1%
	Missing	977	1073	2050
	TOTAL	236	338	574
		100.0%	100.0%	

The DOMES.DUTY question asks the respondents if they are the ones responsible for the general domestic duties—like cleaning, cooking, and washing. What we find is that just 8.1 percent of the men say they do all of the housework themselves compared to 71.3 percent of the women. Most of the men either say they share the housework responsibilities (43.6 percent) or say they do no housework at all (48.3 percent). Only 2.1 percent of the women say they do no housework

The previous table included all married couples in the sample. But, because many more women than men are full-time homemakers, it would seem reasonable that a greater proportion of housework is being done by women. What if we repeat this analysis, but include only those married men and women who are employed full-time?

Data File: **GSS**
Task: **Cross-tabulation**
Row Variable: **69) DOMES.DUTY**
Column Variable: **25) SEX**
➤ Subset Variable: **31) FAM.WORK**
➤ Subset Category: **Include: 3) FULLTIME**
➤ View: **Tables**
➤ Display: **Column %**

DOMES.DUTY   by   SEX
Controls: FAM.WORK: 2 FULLTIME
Cramer's V: 0.693 **

		SEX		
		MALE	FEMALE	TOTAL
DOMES.DUTY	YES,SOLELY	3	80	83
		3.2%	64.0%	37.7%
	YES,EQUALY	57	45	102
		60.0%	36.0%	46.4%
	NO	35	0	35
		36.8%	0.0%	15.9%
	Missing	385	400	785
	TOTAL	95	125	220
		100.0%	100.0%	

Delete the subset variable MARITAL from the previous analysis before continuing.

This table that includes only dual-career couples is not very different from the previous one. Among men whose wives are employed full-time, 60 percent say they share housework responsibilities and 36.8 percent say they do no housework at all. Among married women who are employed full-time, 64 percent say they are solely responsible for the housework. Researchers who have actually tracked the number of hours full-time-employed women spend doing housework have found that, indeed, many women who are employed spend almost as much time doing housework as women who are full-time homemakers. This is why some people refer to employed women's work at home as the *second shift*[1]. It appears that gender roles with regard to providing for the family financially have changed faster than the traditional roles for doing housework.

In the worksheet section that follows, you will have the opportunity to explore some more work and family trends at the global, state, and individual level.

---

[1] Hochschild, Arlie R., with Machung, Anne. *The Second Shift* (New York: Viking Press, 1989).

Marriage and Family

REVIEW QUESTIONS

Based on the first part of this chapter, answer True or False to the following items:

In preindustrial societies, women focused mainly on child care and left subsistence labor to the men.	T	F
Muslim nations have high rates of female employment.	T	F
Although the percentage of women in the paid labor force has increased overall, employment rates for married women have changed very little since 1972.	T	F
In 1920, family incomes were lower in states where a larger percentage of the female population was employed.	T	F
In 1999, median family incomes were lower in states where a larger percentage of the female population was employed.	T	F
Gender role attitudes are a good predictor of whether someone has a one-income or two-income family.	T	F
Those who are in the empty-nest stage are the most likely to have a two-income family.	T	F

EXPLORIT QUESTIONS

1. In the preliminary part of this chapter, you used state-level data to examine the relationship between female employment rates and family income. In your own words, describe how this relationship changed from 1920, to 1940, to 1999.

Chapter 12: Work and Family

2.   Now examine the relationship between per capita income and female employment rates across the nations of the world.

> ➤ *Data File:* **GLOBAL**
> ➤ *Task:* **Scatterplot**
> ➤ *Dependent Variable:* **73) F/M EMPLOY**
> ➤ *Independent Variable:* **22) $ PER CAP**
> ➤ *View:* **Reg. Line**

a. What is the correlation coefficient?   r = _____

b. Are the results statistically significant?   Yes   No

c. Is per capita income a good predictor of which countries are more likely to have a high female employment rate?   Yes   No

d. These results parallel those found in the United States. That is, nations with high levels of female employment also enjoy a high gross national product per capita.   T   F

e. As the decades progress, do you think the relationship between these variables will change? Write a brief paragraph indicating whether or not you think this relationship will change over time and why you think this will be the case. (Hint: You might use the historical pattern found in the United States as the basis for your answer.)

3.   The hypothesis is: **States with higher rates of college graduates have a greater percentage of women in the paid labor force than do other states.**

> ➤ *Data File:* **STATES**
> ➤ *Task:* **Scatterplot**
> ➤ *Dependent Variable:* **41) %FEM.WRK**
> ➤ *Independent Variable:* **56) COLLEGE**
> ➤ *View:* **Reg. Line**

a. What is the value of r for this scatterplot?   r = _____

b. Is r statistically significant?   Yes   No

c. Is the hypothesis supported?   Yes   No

4. The hypothesis is: **States with high rates of births to unwed mothers have lower percentages of women in the paid labor force than do other states.**

> Data File: **STATES**
> Task: **Scatterplot**
> Dependent Variable: **41) %FEM.WRK**
> ➤ Independent Variable: **29) UNWED**
> ➤ View: **Reg. Line**

a. What is the value of r for this scatterplot? r = _____

b. Is r statistically significant? Yes No

c. Is the hypothesis supported? Yes No

d. Based on these results, states with more births to unmarried women tend to have (circle one)
   1. higher female labor force participation rates than other states.
   2. lower female labor force participation rates than other states.
   3. the same female labor force participation rates as other states.

5. Do states with more single females have higher female labor force participation rates than do other states?

> Data File: **STATES**
> Task: **Scatterplot**
> Dependent Variable: **41) %FEM.WRK**
> ➤ Independent Variable: **18) %SINGLE F**
> ➤ View: **Reg. Line**

a. What is the value of r for this scatterplot? r = _____

b. Is r statistically significant? Yes No

c. Based on these results, states with more single females tend to have (circle one)
   1. higher female labor force participation rates than other states.
   2. lower female labor force participation rates than other states.
   3. the same female labor force participation rates as other states.

6. What about states with high percentages of females who are divorced? Are these states likely to have higher rates of female employment?

> Data File: **STATES**
> Task: **Scatterplot**
> Dependent Variable: **41) %FEM.WRK**
> ➤ Independent Variable: **20) %DIV.FEM**
> ➤ View: **Reg. Line**

a. What is the value of r for this scatterplot?   r = _____

b. Is r statistically significant?   Yes   No

c. Based on these results, states with more divorced females tend to have (circle one)
    1. higher female labor force participation rates than other states.
    2. lower female labor force participation rates than other states.
    3. the same female labor force participation rates as other states.

d. Which has the greatest influence on the female employment rate—the proportion of births to unmarried women, the proportion of single females, or the proportion of divorced females? Why might these results occur?

7. Let's return to the GSS to see how a family's work status affects their leisure activities. We'll start with the frequency of socializing with family.

> Data File: **GSS**
>> Task: **Cross-tabulation**
> Row Variable: **78) SOC.KIN**
> Column Variable: **31) FAM.WORK**
>> View: **Tables**
>> Display: **Column %**

a. Is V statistically significant?   Yes   No

b. In comparison to couples where only one spouse is employed full-time, couples where both spouses are employed full-time socialize with relatives (circle one)
    1. more often.
    2. less often.
    3. no more or less often.

8. How about the frequency of socializing with friends?

Data File: **GSS**
Task: **Cross-tabulation**
> Row Variable: **80) SOC.FRIEND**
> Column Variable: **31) FAM.WORK**
>> View: **Tables**
>> Display: **Column %**

a. Is V statistically significant?   Yes   No

b. In comparison to couples where only one spouse is employed full-time, couples where both spouses are employed full-time socialize with friends (circle one)
   1. more often.
   2. less often.
   3. no more or less often.

9. How about dining out at restaurants—are dual-earner families more likely to eat out?

   Data File: **GSS**
   Task: **Cross-tabulation**
   ➤ Row Variable: **85) EAT OUT**
   ➤ Column Variable: **31) FAM.WORK**
   ➤ View: **Tables**
   ➤ Display: **Column %**

   a. Fill in the percentaged results for the *top* row of the table.

	1 FULL TIME	1 FT/ 1 PT	2 FULL TIME
YES	_____%	_____%	_____%

   b. Is V statistically significant?   Yes   No

   c. In comparison to couples where only one spouse is employed full-time, couples where both spouses are employed full-time are (circle one)
      1. more likely to have dined out.
      2. less likely to have dined out.
      3. no more or less likely to have dined out.

10. Finally, we will look at the amount of time spent watching television.

    Data File: **GSS**
    Task: **Cross-tabulation**
    ➤ Row Variable: **82) WATCH TV**
    ➤ Column Variable: **31) FAM.WORK**
    ➤ View: **Tables**
    ➤ Display: **Column %**

    a. Fill in the percentaged results for the *bottom* row of the table.

	1 FULL TIME	1 FT/ 1 PT	2 FULL TIME
4+ HOURS	_____%	_____%	_____%

   b. Is V statistically significant?   Yes   No

   c. In comparison to couples where only one spouse is employed full-time, couples where both spouses are employed full-time are (circle one)

       1.   more likely to watch 4 or more hours of TV per day.

       2.   less likely to watch 4 or more hours of TV per day.

       3.   no more or less likely to watch 4 or more hours of TV per day.

11. Based on the results from Questions 7–10, which leisure activities are affected by a family's work status and which are not? How would you explain these results?

CHAPTER 13

DIVORCE

Tasks: Univariate, Historical Trends, Cross-tabulation, Mapping
Data Files: GSS, TRENDS, STATES, CULTURES

Divorce has become a common experience in America that touches millions of individuals personally and influences the structure and the culture of the communities in which we live. In this chapter, we will explore the prevalence and consequences of divorce from the perspective of history, society, and individuals.

Let's begin by looking at how the divorce rate changed during the twentieth century.

> *Data File:* **TRENDS**
> > *Task:* **Historical Trends**
> > *Variable:* **5) DIV.RATE**

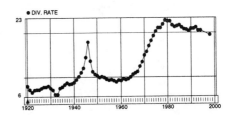

Divorces per 1,000 married females age 15 or older

This graph illustrates changes in divorce rate, the number of divorces per 1,000 married females age 15 or older, from 1920 to 1998. This measure of divorce, known as the *refined divorce rate*, is more accurate than the total number of divorces by itself because it controls for changes in the size of the U.S. population over time.

Overall, the divorce rate was about three times higher in 1998 than it was in 1920. However, the rate of change has not been constant. Most of the increase in the divorce rate since 1950 occurred between 1965 and 1980. Prior to 1965, the divorce rate was increasing at a relatively slow pace.

You also will notice that there was a peak in the divorce rate from 1945 to 1948 immediately following World War II. There are several popular explanations for this spike in the divorce rate. One explanation is that many young people had rushed into marriage during the war years. Once the war was over, these couples found that they were not very compatible and thus divorced. Another explanation is that the period of separation for many couples during the war placed a strain on marriages, thereby increasing the likelihood of divorce. Still another explanation looks at the transformation of gender roles during the war. Many women who previously had not been employed, joined the workforce and managed their households independently during the war years. Some couples may have had difficulty adjusting to these gender-role changes following the war. Regardless of the reason for this peak in the divorce rate, the change was not long lasting. By 1949, the divorce rate had returned to its pre-war level.

The increase in the divorce rate beginning in the 1960s represents a more long-lasting trend. Unlike the increase surrounding World War II, this period of incline lasts for over a decade and is followed by a slight, rather than rapid, decline. Thus, this most recent increase in the divorce rate appears to represent a substantive social and cultural change.

If you ask people who are divorced why their separations occurred, most will probably talk about issues related to personal incompatibility—he was like this, she was like that. A study by Kitson and Sussman (1982) found that the four most common reasons given for divorce were, in descending order of frequency, personality problems, home life, authoritarianism, and differing values. All of these reasons focus on the characteristics of individuals rather than of society. But can we ignore the impact that a couple's social environment may have on their decision to divorce, especially when divorce rates vary from one region of the country to another, or from one society to another? Let's step back and look at divorce from a broader perspective and focus on the nature of the communities in which couples live rather than on the couples themselves. For example, let's examine the geographic distribution of divorce rates across the United States.

> *Data File:* **STATES**
> *Task:* **Mapping**
> *Variable 1:* **13) DIVORCE**
> *View:* **Map**

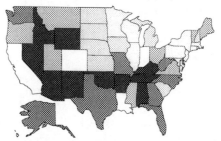

DIVORCE -- 1998: DIVORCES PER 1,000 POPULATION (SA, 2000)

Divorce rates do not vary randomly; there are distinct regional differences. The states with the highest divorce rates are located in the West and the South. Let's look at the actual rankings.

Data File: **STATES**
Task: **Mapping**
Variable 1: **13) DIVORCE**
> *View:* **List: Rank**

DIVORCE: 1998: Divorces per 1,000 population

RANK	CASE NAME	VALUE
1	Nevada	6.8
2	Wyoming	6.7
3	Oklahoma	6.5
4	New Mexico	6.5
5	Arkansas	6.3
6	Tennessee	6.0
7	Alabama	5.91
8	Arizona	5.8
9	Kentucky	5.8
10	Idaho	5.4

Here we see that Nevada has the highest divorce rate with 6.8 divorces for every 1,000 residents living there. Wyoming is second, followed by Oklahoma, New Mexico, and Arkansas. Massachusetts has the lowest divorce rate (2.2 divorces per 1,000 residents) followed by Maryland, New Jersey, and New York.

Have western and southern states always had the highest divorce rates?

Reference

Kitson, Gay, and Marvin Sussman. "Marital Complaints, Demographic Characteristics, and Symptoms of Mental Distress in Divorce." *Journal of Marriage and the Family*, 44:1 (1982) 87–101.

Data File: **STATES**
Task: **Mapping**
Variable 1: **13) DIVORCE**
➤ Variable 2: **72) DIVORCE 22**
➤ Views: **Map**

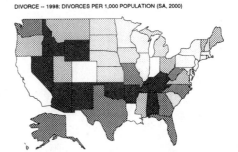

r = 0.514**

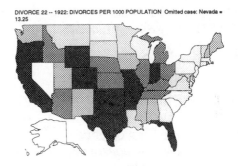

The maps are very similar, and the correlation coefficient indicates a significant positive relationship of 0.514**. The states that had higher divorce rates in 1998 also had higher divorce rates in 1922. These findings are interesting because they demonstrate that we must go beyond *personal incompatibility* in order to explain divorce. If divorce rates were *entirely* dependent on decisions made by autonomous individuals, divorce rates would be very unpredictable. States that had a high divorce rate one year would be no more or less likely to have a high divorce rate the following year. But the comparison of the divorce rates of 1922 with those of 1998 shows very little fluctuation. There must be something about living in the West or the South that increases the likelihood of divorce.

Let's compare the map of divorce rates with other maps to see if we can identify some possible reasons why some states have a higher divorce rate than others. We can begin with education. A more educated population probably has the economic resources and/or the interpersonal skills that increase marital stability.

Data File: **STATES**
Task: **Mapping**
Variable 1: **13) DIVORCE**
➤ Variable 2: **56) COLLEGE**
➤ Views: **Map**

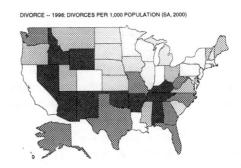

r = –0.575**

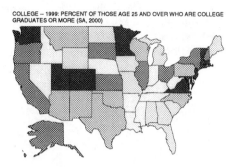

COLLEGE -- 1999: PERCENT OF THOSE AGE 25 AND OVER WHO ARE COLLEGE GRADUATES OR MORE (SA, 2000)

These maps are nearly mirror images of one another, as indicated by the *negative* relationship of –0.575**. The more college graduates there are in a state, the lower the divorce rate.

Another regional characteristic that may affect the divorce rate is population growth.

Data File: **STATES**
Task: **Mapping**
Variable 1: **13) DIVORCE**
➤ *Variable 2:* **6) POP.GROW**
➤ *Views:* **Map**

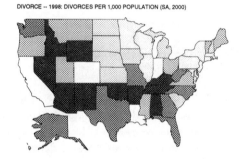

DIVORCE -- 1998: DIVORCES PER 1,000 POPULATION (SA, 2000)

r = 0.500**

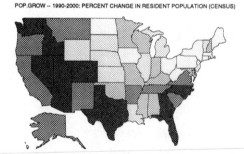

POP.GROW -- 1990-2000: PERCENT CHANGE IN RESIDENT POPULATION (CENSUS)

These two maps are very similar, and there is a *positive* correlation coefficient (0.500**). States that are experiencing the greatest population growth also have higher divorce rates. This variable is a good indicator of social integration. People are generally more integrated into a community when there are fewer of the disruptions that result from people moving in and out. So, how does social integration influence the divorce rate? One explanation is that long-term relationships with people in the community may diffuse some of the responsibility placed on marital relationships, thus making it easier for

Marriage and Family

the marriage to survive. There are people around to meet the needs that spouses may not be able to meet by themselves. Another explanation is that it is easier for couples to get divorced in communities where they feel more anonymous. There is more pressure to stay together in communities where everyone knows everyone else.

Let's switch to a more individual level perspective to see if we can find some similar divorce-related trends. We'll start by looking at the current marital status of the entire GSS sample.

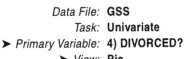

> Data File: **GSS**
> Task: **Univariate**
> Primary Variable: **1) MARITAL**
> View: **Pie**

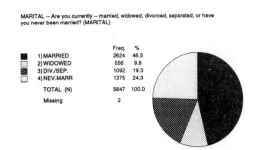

MARITAL -- Are you currently -- married, widowed, divorced, separated, or have you never been married? (MARITAL)

	Freq.	%
1) MARRIED	2624	46.5
2) WIDOWED	556	9.8
3) DIV./SEP.	1092	19.3
4) NEV.MARR	1375	24.3
TOTAL (N)	5647	100.0
Missing	2	

Just under one-half of the GSS respondents are currently married. Those who are currently divorced make up the third largest category, at 19.3 percent. Of course, some of those who are currently married may have been divorced previously. Let's see how many people who have ever been married have been divorced at some time.

Data File: **GSS**
Task: **Univariate**
> Primary Variable: **4) DIVORCED?**
> View: **Pie**

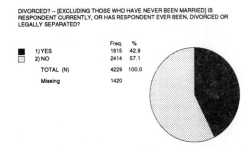

DIVORCED? -- [EXCLUDING THOSE WHO HAVE NEVER BEEN MARRIED] IS RESPONDENT CURRENTLY, OR HAS RESPONDENT EVER BEEN, DIVORCED OR LEGALLY SEPARATED?

	Freq.	%
1) YES	1815	42.9
2) NO	2414	57.1
TOTAL (N)	4229	100.0
Missing	1420	

Forty-three percent of those who have ever been married have been divorced at some time. Undoubtedly you have heard or read that one-half of all marriages end in divorce—so why is the number of divorced persons in this sample less than that? First, we must remember how the popular "50 percent" statistic is computed. It is derived by dividing the number of divorces each year by the number of marriages. Recently, there has been one divorce for every two marriages—thus, one-half of all marriages end in divorce. However, what this figure does not reflect is that some people marry and divorce more than once. Thus, although one-half of all marriages may end in divorce, less than one-half of all married people have ever been divorced.

Using the state-level data, we found that a state's divorce rate was negatively correlated with the percentage of college graduates. Furthering one's education will ultimately benefit a marriage by providing financial security. The college experience also develops other skills that may enhance a marriage relationship, such as improved communication and an appreciation of other people's ways of seeing the world. Therefore, the hypothesis is: **Those who have attended college will be less likely to have divorced.**

Data File: **GSS**
➤ Task: **Cross-tabulation**
➤ Row Variable: **4) DIVORCED?**
➤ Column Variable: **27) EDUCATION**
➤ Control Variable: **25) SEX**
➤ View: **Tables (MALE)**
➤ Display: **Column %**

DIVORCED?   by   EDUCATION
Controls: SEX: MALE
Cramer's V: 0.066 *

		EDUCATION				
		NO HS GRAD	HS GRAD	COLL EDUC	Missing	TOTAL
D I V O R C E D ?	YES	162	218	380	2	760
		49.5%	45.0%	41.0%		43.7%
	NO	165	266	547	6	978
		50.5%	55.0%	59.0%		56.3%
	Missing	100	197	417	1	715
	TOTAL	327	484	927	9	1738
		100.0%	100.0%	100.0%		

The results in this table, which includes only males, are consistent with the hypothesis. Reading across the top of the table, we find 49.5 percent of males who dropped out of high school have ever been divorced compared to 41 percent for those who went to college. The relationship is not very strong, but it is statistically significant (V = 0.066*).

Let's look at the results for women.

Data File: **GSS**
Task: **Cross-tabulation**
Row Variable: **4) DIVORCED?**
Column Variable: **27) EDUCATION**
Control Variable: **25) SEX**
➤ View: **Tables (FEMALE)**
➤ Display: **Column %**

DIVORCED?   by   EDUCATION
Controls: SEX: FEMALE
Cramer's V: 0.027

		EDUCATION				
		NO HS GRAD	HS GRAD	COLL EDUC	Missing	TOTAL
D I V O R C E D ?	YES	192	339	514	8	1045
		45.0%	41.1%	42.1%		42.3%
	NO	235	486	707	2	1428
		55.0%	58.9%	57.9%		57.7%
	Missing	122	168	413	2	705
	TOTAL	427	825	1221	12	2473
		100.0%	100.0%	100.0%		

The percentages in this table are much closer. There is only a three percentage-point difference between those who did not graduate from high school (45 percent) and those who went to college (42.1 percent). For women, the relationship between education and having ever been divorced is not statistically significant (V = 0.027).

Why would having a college education decrease the likelihood of divorce for men but not for women? One explanation would be that the positive effect a college education can have on a marriage with regard to family income and social skills may be offset by the fact that women with a college education have more alternatives to marriage with regard to financial stability and career opportunities.

Next, let's look at the impact of children on the likelihood of divorce. Do parents stay together for the sake of their children?

Data File:	**GSS**
Task:	**Cross-tabulation**
Row Variable:	**4) DIVORCED?**
➤ Column Variable:	**13) # CHILDREN**
➤ View:	**Tables**
➤ Display:	**Column %**

DIVORCED?   by   # CHILDREN
Cramer's V: 0.012

# CHILDREN					
		NONE	1-2	3 AND UP	Missing	TOTAL
YES		260	911	634	10	1805
		43.6%	43.2%	42.1%		42.8%
NO		336	1200	872	6	2408
		56.4%	56.8%	57.9%		57.2%
Missing		1019	280	114	7	1420
TOTAL		596	2111	1506	23	4213
		100.0%	100.0%	100.0%		

Remember to remove the control variable before continuing.

The difference between those who have children and those who do not is very slight. There are no asterisks after the V of 0.012, which means the difference is not statistically significant. It appears that those marriages that are headed toward divorce will end up there whether or not children are involved. If children do not prevent divorce, do they at least slow down the process? Maybe parents will wait at least until their children are older before they separate. Let's test this idea by looking at a cross-tabulation between those who are currently divorced and the number of household members who are under age 6.

Data File:	**GSS**
Task:	**Cross-tabulation**
Row Variable:	**4) DIVORCED?**
➤ Column Variable:	**17) KIDS<6**
➤ View:	**Tables**
➤ Display:	**Column %**

DIVORCED?   by   KIDS<6
Cramer's V: 0.125 **

KIDS<6					
		NO	YES	Missing	TOTAL
YES		1595	208	12	1803
		45.8%	29.3%		43.0%
NO		1885	502	27	2387
		54.2%	70.7%		57.0%
Missing		1243	164	13	1420
TOTAL		3480	710	52	4190
		100.0%	100.0%		

The difference here is a bit larger. Among those who have no preschool-age children, 45.8 percent have been divorced, compared to 29.3 percent for those who have one or more children under age 6. This difference is statistically significant (V = 0.125**). This finding supports what is sometimes called the braking hypothesis, which states that children do not prevent divorce but that they do slow down the process.

The timing of when children are born may also impact the likelihood of divorce. Because age tends to bring about financial and emotional security, we'll hypothesize that among respondents who have children, **those who waited until they were older to have children will be less likely to have been divorced.**

Data File:	**GSS**
Task:	**Cross-tabulation**
Row Variable:	**4) DIVORCED?**
➤ Column Variable:	**18) AGE KD BRN**
➤ View:	**Tables**
➤ Display:	**Column %**

DIVORCED?   by   AGE KD BRN
Cramer's V: 0.155 **

	AGE KD BRN				
		<20	20-29	30 AND UP	Missing	TOTAL
DIVORCED?	YES	444	917	177	277	1538
		57.0%	40.1%	34.3%		42.9%
	NO	335	1371	339	369	2045
		43.0%	59.9%	65.7%		57.1%
	Missing	173	180	28	1039	1420
	TOTAL	779	2288	516	1685	3583
		100.0%	100.0%	100.0%		

Here we see that those who had a child before reaching age 20 (57 percent) are much more likely than those who waited until their twenties (40.1 percent) or later (34.3 percent) to have ever been divorced. Having a child at a young age greatly increases the likelihood of divorce (V = 0.155**).

Now let's look at religion. Does church attendance influence the probability of divorce?

Data File:	**GSS**
Task:	**Cross-tabulation**
Row Variable:	**4) DIVORCED?**
➤ Column Variable:	**50) ATTEND**
➤ View:	**Tables**
➤ Display:	**Column %**

DIVORCED?   by   ATTEND
Cramer's V: 0.179 **

	ATTEND				
		NEVER	MONTH/YRLY	WEEKLY	Missing	TOTAL
DIVORCED?	YES	418	917	447	33	1782
		53.3%	47.8%	31.2%		43.1%
	NO	366	1001	984	63	2351
		46.7%	52.2%	68.8%		56.9%
	Missing	345	769	278	28	1420
	TOTAL	784	1918	1431	124	4133
		100.0%	100.0%	100.0%		

Overall, those who attend church at least weekly (31.2 percent) are less likely than those who attend less often (47.8 percent) or not at all (53.3 percent) to have ever been divorced. Of course, we don't know if weekly church attendance came before or after the divorce. It is possible that some people began attending church regularly after they got divorced. However, given the strength of the relationship (V = 0.179**) and the fact that most churches discourage the dissolution of marriages except for special circumstances, it seems likely that those who attend church regularly are less likely to get divorced.

Finally, let's see if those whose parents were divorced when they were growing up are more likely to have ever been divorced themselves.

Data File:	**GSS**
Task:	**Cross-tabulation**
Row Variable:	**4) DIVORCED?**
➤ Column Variable:	**45) FAMILY @16**
➤ View:	**Tables**
➤ Display:	**Column %**

DIVORCED?   by   FAMILY @16
Cramer's V: 0.098 **

	FAMILY @16		
		YES	NO	TOTAL
DIVORCED?	YES	1209	606	1815
		39.9%	50.7%	42.9%
	NO	1824	590	2414
		60.1%	49.3%	57.1%
	Missing	862	558	1420
	TOTAL	3033	1196	4229
		100.0%	100.0%	

Here we see that those whose parents were divorced when the respondent was 16 years old (50.7 percent) are more likely than those whose parents were together (39.9 percent) to have ever been divorced themselves (V = 0.098**). Some of this relationship may be the effect of modeling—those whose parents were divorced may be more likely to view divorce as an option for themselves. However, being raised by parents who are divorced may be associated with other socioeconomic factors that may increase the likelihood of divorce. For example, are those whose parents were divorced less likely to have graduated high school or gone to college?

Data File: **GSS**

Task: **Cross-tabulation**

➤ Row Variable: **27) EDUCATION**

➤ Column Variable: **45) FAMILY @16**

➤ View: **Tables**

➤ Display: **Column %**

EDUCATION by FAMILY @16
Cramer's V: 0.130 **

		FAMILY @16		
		YES	NO	TOTAL
EDUCATION	NO HS GRAD	555	421	976
		14.3%	24.1%	17.3%
	HS GRAD	1141	533	1674
		29.4%	30.5%	29.7%
	COLL EDUC	2186	792	2978
		56.3%	45.4%	52.9%
	Missing	13	8	21
	TOTAL	3882	1746	5628
		100.0%	100.0%	

Looking at these results we see that those whose parents were divorced (24.1 percent) are more likely than those whose parents were together (14.3 percent) to have less than a high school education. Conversely, those whose parents were together (56.3 percent) are more likely than those whose parents were divorced (45.4 percent) to have gone to college. The results are statistically significant (V = 0.130**).

Because education and income are correlated, we would expect to see similar results when we look at personal incomes.

Data File: **GSS**

Task: **Cross-tabulation**

➤ Row Variable: **29) R.INCOME**

➤ Column Variable: **45) FAMILY @16**

➤ View: **Tables**

➤ Display: **Column %**

R.INCOME by FAMILY @16
Cramer's V: 0.113 **

		FAMILY @16		
		YES	NO	TOTAL
R.INCOME	$0K-17.4K	773	454	1227
		30.5%	40.1%	33.5%
	17.5K-34.9	891	401	1292
		35.1%	35.5%	35.2%
	$35K +	872	276	1148
		34.4%	24.4%	31.3%
	Missing	1359	623	1982
	TOTAL	2536	1131	3667
		100.0%	100.0%	

As expected, these results are very similar to the results in the previous table. Those who were raised in a two-parent family are more likely to have higher personal incomes (V = 0.113**).

Now that we know that those whose parents were divorced tend to have lower levels of education and income, let's return to the relationship between 45) FAMILY @16 and 4) DIVORCED?. This time, we will include a control variable for personal income.

<table>
<tr><td>

Data File: **GSS**

Task: **Cross-tabulation**

➤ *Row Variable:* **4) DIVORCED?**

➤ *Column Variable:* **45) FAMILY @16**

➤ *Control Variable:* **29) R.INCOME**

➤ *View:* **Tables ($0K–17.4K)**

➤ *Display:* **Column %**

</td></tr>
</table>

DIVORCED?   by   FAMILY @16
Controls: R.INCOME: $0K–17.4K
Cramer's V: 0.145 **

		FAMILY @16		
		YES	NO	TOTAL
DIVORCED?	YES	212	143	355
		40.5%	55.9%	45.5%
	NO	312	113	425
		59.5%	44.1%	54.5%
	Missing	249	198	447
	TOTAL	524	256	780
		100.0%	100.0%	

The results for this first table are similar to the earlier results. Among those in the lowest income category, respondents whose parents were divorced are more likely to have been divorced themselves (V = 0.145**). Let's look at the results for the next income category.

Data File: **GSS**

Task: **Cross-tabulation**

Row Variable: **4) DIVORCED?**

Column Variable: **45) FAMILY @16**

Control Variable: **29) R.INCOME**

➤ *View:* **Tables ($17.5K–34.9K)**

➤ *Display:* **Column %**

DIVORCED?   by   FAMILY @16
Controls: R.INCOME: 17.5K–34.9
Cramer's V: 0.058

		FAMILY @16		
		YES	NO	TOTAL
DIVORCED?	YES	316	146	462
		47.9%	54.3%	49.7%
	NO	344	123	467
		52.1%	45.7%	50.3%
	Missing	231	132	363
	TOTAL	660	269	929
		100.0%	100.0%	

The percentages for those in this middle income category are much closer and are not statistically significant (V = 0.058).

Data File: **GSS**

Task: **Cross-tabulation**

Row Variable: **4) DIVORCED?**

Column Variable: **45) FAMILY @16**

Control Variable: **29) R.INCOME**

➤ *View:* **Tables ($35K +)**

➤ *Display:* **Column %**

DIVORCED?   by   FAMILY @16
Controls: R.INCOME: $35K +
Cramer's V: 0.032

		FAMILY @16		
		YES	NO	TOTAL
DIVORCED?	YES	306	102	408
		43.7%	47.4%	44.6%
	NO	394	113	507
		56.3%	52.6%	55.4%
	Missing	172	61	233
	TOTAL	700	215	915
		100.0%	100.0%	

The percentages for those in the highest income category are also quite close, and again Cramer's V is not statistically significant (V = 0.032). Thus, the higher divorce rate among those whose parents were divorced is partially explained by socioeconomic factors related to divorce rather than just modeling by the divorced parents.

So far, we have been looking at the factors that may increase the likelihood of divorce. In the worksheet section that follows, you will be asked to analyze lifestyle differences between those who are currently divorced and those who are currently married. And, finally, you will have the opportunity to look at some divorce-related variables from the CULTURES data set.

Marriage and Family

Workbook exercises and software are copyrighted. Copying is prohibited by law.

REVIEW QUESTIONS

Based on the first part of this exercise, answer True or False to the following items:

Divorce rates in the United States hit their peak in 1998.	T	F
The states with the highest divorce rates are in the West and the South.	T	F
The geographic distribution of divorce rates in the United States has changed considerably since 1922.	T	F
States with more college graduates tend to have higher divorce rates.	T	F
States with fast-growing populations tend to have lower divorce rates.	T	F
Around 40 percent of those surveyed in the GSS have been divorced at some time.	T	F
Parents with children under 6 years of age are less likely to be currently divorced.	T	F
Those who waited until age 30 to have their first child are more likely to have been divorced.	T	F

EXPLORIT QUESTIONS

1. Women who are currently divorced likely understand the necessity of being able to obtain employment. So, the hypothesis is: **Women who are currently divorced will be more likely to approve of married women being employed.**

    > ➤ *Data File:* **GSS**
    > ➤ *Task:* **Cross-tabulation**
    > ➤ *Row Variable:* **68) MARR.ROLES**
    > ➤ *Column Variable:* **6) DIV/MAR**
    > ➤ *Subset Variable:* **25) SEX**
    > ➤ *Subset Category:* **Include: 2) Female**
    > ➤ *View:* **Tables**
    > ➤ *Display:* **Column %**

    a. Fill in the percentaged results for the *bottom* row of the table.

	DIVORCED	MARRIED
DISAGREE	_____%	_____%

b.   What is the value of V?   V = _____

c.   Is V statistically significant?   Yes   No

d.   Is the hypothesis supported?   Yes   No

e.   Women who are currently divorced have less traditional sex-role attitudes.   T   F

2.   What is the relationship between marital status and personal happiness? (Note: Use the [Clear All] button to remove the subset variable that was previously selected.)

> Data File: **GSS**
> Task: **Cross-tabulation**
> ➤ Row Variable: **73) HAPPY?**
> ➤ Column Variable: **6) DIV/MAR**
> ➤ View: **Tables**
> ➤ Display: **Column %**

a.   Fill in the percentaged results for the *top* row of the table.

	DIVORCED	MARRIED
VERY HAPPY	_____%	_____%

b.   What is the value of V?   V = _____

c.   Is V statistically significant?   Yes   No

d.   Are those who are divorced less likely to be very happy?   Yes   No

3.   What is the relationship between marital status and personal health?

> Data File: **GSS**
> Task: **Cross-tabulation**
> ➤ Row Variable: **76) HEALTH**
> ➤ Column Variable: **6) DIV/MAR**
> ➤ View: **Tables**
> ➤ Display: **Column %**

a.   Fill in the percentaged results for the *top* row of this table.

	DIVORCED	MARRIED
EXCELLENT	_____%	_____%

b.   What is the value of V?   V = _____

   c. Is V statistically significant?   Yes   No

   d. Are married people in better physical health than divorced people?   Yes   No

   e. Explain how you think health and the status of being married or divorced are related.

4.   Finally, what is the relationship between marital status and family income?

> *Data File:* **GSS**
> *Task:* **Cross-tabulation**
> ➤ *Row Variable:* **28) INCOME**
> ➤ *Column Variable:* **6) DIV/MAR**
> ➤ *View:* **Tables**
> ➤ *Display:* **Column %**

   a. What percentage of those who are married are in the lowest income category?   _____%

   b. What percentage of those who are divorced are in the lowest income category?   _____%

   c. What percentage of those who are married are in the highest income category?   _____%

   d. What percentage of those who are divorced are in the highest income category?   _____%

   e. How do you think income and the status of being divorced or married are related?

5.   The CULTURES data file has a variable named DIV COMMON that allows us to examine how common divorce was in preindustrial societies. Unfortunately, when this variable is used in a cross-tabulation, fewer than 50 societies are generally coded on both variables. This results in very few analyses being statistically significant. However, for the following examples, focus on the general trends that appear in the tables, rather than the statistical significance. Even results that do not meet scientific guidelines can be of interest.

➤ *Data File:* **CULTURES**
➤ *Task:* **Cross-tabulation**
➤ *Row Variable:* **27) DIV COMMON**
➤ *Column Variable:* **52) FEM.PROPTY**
➤ *View:* **Tables**
➤ *Display:* **Column %**

a. Fill in the *bottom* row of this table.

	LOW	MEDIUM	HIGH
COMMON	_____%	_____%	_____%

b. Does the degree to which women have control over property seem to be related to whether divorce is common in a society?   Yes   No

c. Divorce is most common in societies in which women have high levels of control over property.   T   F

6. Repeat the previous analysis using 22) PLAY MATES as the column variable. (PLAY MATES is the amount of leisure time that husbands and wives spend together.)

a. Fill in the *bottom* row of this table.

	LITTLE	SOME	MUCH
COMMON	_____%	_____%	_____%

b. Does the degree to which spouses spend their leisure time together seem to be related to whether divorce is common in a society?   Yes   No

c. Based on these results, does it seem reasonable to conclude that couples who play together, stay together?   Yes   No

7. Repeat the previous analysis using 21) SPOUSE EAT as the column variable. (SPOUSE EAT is whether husbands and wives usually eat meals together.)

a. Fill in the *bottom* row of this table.

	NO	YES
COMMON	_____%	_____%

b. Does eating with a spouse seem to be related to whether divorce is common in a society?   Yes   No

   c. Based on these results, does it seem reasonable to suggest that couples who eat together, stay together?   Yes   No

8.   Repeat the previous analysis using 19) COUPLE BED as the column variable. (COUPLE BED is whether husbands and wives sleep together in the same bed.)

   a. Fill in the *bottom* row of this table.

	SAME BED	APART
COMMON	_____%	_____%

   b. Does sleeping together in the same bed seem to be related to whether divorce is common in a society?   Yes   No

   c. Based on these results, does it seem reasonable to suggest that couples who sleep together, stay together?   Yes   No

CHAPTER **14**

REMARRIAGE

Tasks: Univariate, Historical Trends, Cross-tabulation
Data Files: GSS, TRENDS

D uring the past few decades, there has been an increase in what social scientists refer to as serial monogamy. In contrast to monogamy, which is marriage to only one partner, serial monogamy refers to remarriage after a divorce. The purpose of this chapter will be to examine the frequency and characteristics of remarriages.

Let's begin by looking at how often remarriage occurs following a divorce.

> ➤ Data File: **GSS**
> ➤ Task: **Univariate**
> ➤ Primary Variable: **8) DIV/REMAR.**
> ➤ View: **Pie**

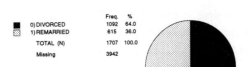

DIV/REMAR. -- IF EVER DIVORCED, IS RESPONDENT REMARRIED?

		Freq.	%
■	0) DIVORCED	1092	64.0
▨	1) REMARRIED	615	36.0
	TOTAL (N)	1707	100.0
	Missing	3942	

This graph represents the current marital status of individuals who have ever been divorced. Thirty-six percent of the respondents are remarried. Of course, some of those who are included in the divorce category may have actually been remarried at some time, and then divorced again. So, the actual number of divorced respondents who have ever been remarried is probably higher than what is shown. Keeping that in mind, let's see how this percentage compares to previous years of the General Social Survey.

> ➤ Data File: **TRENDS**
> ➤ Task: **Historical Trends**
> ➤ Variable: **10) REMARRIED%**

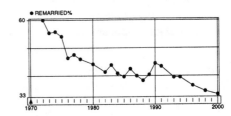

Percentage of respondents who have been divorced or separated and are
currently remarried

227

The percentage of divorced respondents who were remarried at the time they were surveyed has steadily declined in the past several decades. In 1972, 60 percent of those who divorced were remarried compared to just over 33 percent in 2000. Among those surveyed, there are fewer people in remarriages now than in the past.

Why are fewer divorced people remarried? Part of the explanation for this phenomenon can be found by looking at the employment rate of married females during this time span.

Data File: **TRENDS**
Task: **Historical Trends**
➤ Variables: **10) REMARRIED%**
27) M FEM EMP%

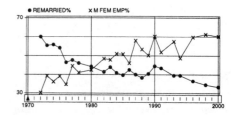

Percent of respondents who have been divorced or separated and are currently remarried

Percent of married females employed part time or full time

As the married female employment rate rises, the remarriage rate declines. In the past, when most married women were not employed, remarriage was much more of an economic necessity than it is today when most women are employed before they get divorced and thus can support themselves.

Regardless of the recent declines in the remarriage rate, over one-third of those who divorced have already remarried. National studies have found that eventually three-fourths of those who divorce eventually remarry (National Center for Health Statistics, Advance Data, May 2001). The rising divorce rate does not necessarily mean Americans are disillusioned with the institution of marriage. The disappointment generally appears to be with particular marriage relationships.

How do remarriages compare with first marriages? One could argue that remarriages will be happier, since people should be more aware of what they are looking for in a mate and have found someone who meets their current needs. On the other hand, first marriages may be happier because they have navigated the transitions that have pulled other relationships apart. Let's see which argument is supported.

➤ Data File: **GSS**
➤ Task: **Cross-tabulation**
➤ Row Variable: **74) HAP.MARR.?**
➤ Column Variable: **9) REMARRIAGE**
➤ View: **Tables**
➤ Display: **Column %**

HAP.MARR.?   by   REMARRIAGE
Cramer's V: 0.026

		REMARRIAGE			
		1 MARRIAGE	REMARRIAGE	Missing	TOTAL
HAP.MARR.?	VERY HAPPY	1247	382	15	1629
		63.3%	62.7%		63.1%
	PRET.HAPPY	670	204	5	874
		34.0%	33.5%		33.9%
	NOT TOO	54	23	1	77
		2.7%	3.8%		3.0%
	Missing	14	6	3028	3048
	TOTAL	1971	609	3049	2580
		100.0%	100.0%		

It appears that neither argument is correct—the percentages of those who have had one marriage are not very different from the percentages of those who are remarried. The slight differences you do see are not statistically significant (V = 0.026). Remarriages are no more or less happy than first marriages.

Perhaps there are other differences that we can identify. For example, because divorce is a departure from traditional norms, we could hypothesize that those who are remarried will have less traditional gender-role attitudes than those who are in their first marriages.

Data File: **GSS**
Task: **Cross-tabulation**
➤ Row Variable: **68) MARR.ROLES**
➤ Column Variable: **9) REMARRIAGE**
➤ View: **Tables**
➤ Display: **Column %**

MARR.ROLES   by   REMARRIAGE
Cramer's V: 0.068

		REMARRIAGE			
		1 MARRIAGE	REMARRIAGE	Missing	TOTAL
MARR.ROLES	AGREE	101	40	133	141
		23.3%	27.2%		24.3%
	NEITHER	97	24	116	121
		22.4%	16.3%		20.9%
	DISAGREE	235	83	407	318
		54.3%	56.5%		54.8%
	Missing	1552	468	2393	4413
	TOTAL	433	147	3049	580
		100.0%	100.0%		

Again, the percentages are very similar. Persons who have divorced and remarried are no more or less likely than those who have never divorced to believe that it is the husband's job to earn money and the wife's job to take care of the house and the family (V = 0.068). Although remarriages vary from the traditional norm of marrying for life, people who remarry can be just as traditional as people who have been married only once. People carry their traditional gender-role attitudes with them from one marriage relationship to another.

Although gender-role attitudes may not be different in remarriages, are there differences in actual role enactment? For example, are women who are remarried more likely to be employed?

Data File: **GSS**
Task: **Cross-tabulation**
➤ Row Variable: **32) MAR.F.WRK**
➤ Column Variable: **9) REMARRIAGE**
➤ View: **Tables**
➤ Display: **Column %**

MAR.F.WRK   by   REMARRIAGE
Cramer's V: 0.081 *

		REMARRIAGE			
		1 MARRIAGE	REMARRIAGE	Missing	TOTAL
MAR.F.WRK	FULL TIME	476	165	5	641
		50.3%	59.6%		52.4%
	PART-TIME	165	44	2	209
		17.4%	15.9%		17.1%
	KEEP HOUSE	305	68	1	373
		32.2%	24.5%		30.5%
	Missing	1039	338	3041	4418
	TOTAL	946	277	3049	1223
		100.0%	100.0%		

Women who are remarried are more likely to be employed, with 59.6 percent of women in a remarriage being employed full-time compared to 50.3 percent of women in their first marriage. There is a similar difference in the bottom row of the table with 32.2 percent of women in their first marriage keeping house full-time compared to 24.5 percent of women who have been married before. These differences are not great, but they are statistically significant (V = 0.081*).

Let's look now at the structure of the household for first marriages versus remarriages. Because remarriages often involve children from the previous marriages of one, or both, spouses, we might hypothesize that persons who are remarried will have more people living in their households.

Data File: **GSS**

Task: **Cross-tabulation**

➤ Row Variable: **15) HH SIZE**

➤ Column Variable: **9) REMARRIAGE**

➤ View: **Tables**

➤ Display: **Column %**

HH SIZE   by   REMARRIAGE
Cramer's V: 0.069 **

		REMARRIAGE			
		1 MARRIAGE	REMARRIAGE	Missing	TOTAL
HH SIZE	2	828	304	734	1132
		42.3%	50.2%		44.1%
	3-4	849	233	687	1082
		43.4%	38.4%		42.2%
	5 OR MORE	281	69	142	350
		14.4%	11.4%		13.7%
	Missing	27	9	1486	1522
	TOTAL	1958	606	3049	2564
		100.0%	100.0%		

Contrary to our hypothesis, there do not appear to be any substantive differences in this table. In fact, households that result from a remarriage (50.2 percent) are somewhat more likely than the households of couples in their first marriage (42.3 percent) to be living by themselves (V = 0.069**). There is very little difference in any of the other categories of household size. Although some reconstituted households may involve children from one or more previous marriages, overall, remarriages do not lead to larger households.

Divorce and remarriage will inevitably lead to changes in extended family relationships as new people are introduced into the family system. But do these changes necessarily influence how often people socialize with members of their extended family?

Data File: **GSS**

Task: **Cross-tabulation**

➤ Row Variable: **78) SOC.KIN**

➤ Column Variable: **9) REMARRIAGE**

➤ View: **Tables**

➤ Display: **Column %**

SOC.KIN   by   REMARRIAGE
Cramer's V: 0.044

		REMARRIAGE			
		1 MARRIAGE	REMARRIAGE	Missing	TOTAL
SOC.KIN	DAILY/WKLY	741	212	1059	953
		56.3%	51.2%		55.1%
	MONTH/YEAR	543	190	815	733
		41.3%	45.9%		42.4%
	NEVER	32	12	124	44
		2.4%	2.9%		2.5%
	Missing	669	201	1051	1921
	TOTAL	1316	414	3049	1730
		100.0%	100.0%		

Overall, those who are remarried socialize with kin about as frequently as those who are in their first marriage. The slight differences between the two groups are not statistically significant (V = 0.044).

So far, we have compared remarriages with first marriages and have found that remarriages are not very different from first marriages with regard to marital happiness, gender-role attitudes, household size, or the frequency of socializing with extended family. The only statistically significant difference we found was that women who are remarried are more likely to be employed.

In the exercises that follow, we will turn our attention to comparing those who have remarried following a divorce with those who have not remarried.

WORKSHEET

NAME: _____

COURSE: _____

DATE: _____

CHAPTER 14

REVIEW QUESTIONS

Based on the first part of this chapter, answer True or False to the following items:

Serial monogamy refers to the marriage pattern in which people are married more than once.	T	F
According to the National Center for Health Statistics, about three-fourths of all divorced Americans will remarry at some time.	T	F
Statistically speaking, remarriages are as happy as first marriages.	T	F
Those who are remarried tend to have less traditional gender-role attitudes than those who are in their first marriage.	T	F
Women who have remarried are more likely to be employed.	T	F
Those who are remarried tend to have more people living in their households.	T	F
In comparison to people who have remarried, people who are divorced socialize more frequently with extended family.	T	F

EXPLORIT QUESTIONS

1. Who is more likely to remarry after divorce—a man or a woman?

    ➤ *Data File:* **GSS**
    ➤ *Task:* **Cross-tabulation**
    ➤ *Row Variable:* **8) DIV/REMAR.**
    ➤ *Column Variable:* **25) SEX**
    ➤ *View:* **Tables**
    ➤ *Display:* **Column %**

    a. Fill in the percentaged results for the *bottom* row of the table.

	MALE	FEMALE
REMARRIED	_____%	_____%

    b. What is the value of V?   V = _____

c. Is V statistically significant?   Yes   No

d. Men are more likely than women to remarry.   T   F

e. Explain why you think these results occur.

2. The hypothesis is: **Those who believe that men should be the economic provider for the family while women take care of the home will be more likely to remarry.**

> Data File: **GSS**
> Task: **Cross-tabulation**
> Row Variable: **8) DIV/REMAR.**
> ➤ Column Variable: **68) MARR.ROLES**
> ➤ View: **Tables**
> ➤ Display: **Column %**

a. Fill in the percentaged results for the *bottom* row of the table.

	AGREE	**DISAGREE**	**NEITHER**
REMARRIED	_____%	_____%	_____%

b. What is the value of V?   V = _____

c. Is V statistically significant?   Yes   No

d. Is the hypothesis supported?   Yes   No

e. How would you explain these results?

3.   Does a respondent's income influence the likelihood of remarriage? If so, is the effect the same for women as it is for men? To answer this last question we will need to include sex as a control variable.

> Data File: **GSS**
> Task: **Cross-tabulation**
> Row Variable: **8) DIV/REMAR.**
> ➤ Column Variable: **29) R.INCOME**
> ➤ Control Variable: **25) SEX**
> ➤ View: **Tables (MALE)**
> ➤ Display: **Column %**

a.   What percentage of men in the highest income category are remarried?   _____%

b.   What percentage of men in the lowest income category are remarried?   _____%

c.   Is V statistically significant?   Yes   No

Click on the arrow to advance to the table for females.

d.   What percentage of women in the highest income category are remarried?   _____%

e.   What percentage of women in the lowest income category are remarried?   _____%

f.   Is V statistically significant?   Yes   No

g.   Summarize how a respondent's income influences the likelihood of marriage for men and for women. How would you explain these results?

4.   Now let's see if the likelihood of remarriage is related to race. (Don't forget to clear the control variable from the previous question.)

> Data File: **GSS**
> Task: **Cross-tabulation**
> Row Variable: **8) DIV/REMAR.**
> ➤ Column Variable: **26) RACE**
> ➤ View: **Tables**
> ➤ Display: **Column %**

a. Fill in the percentaged results for the *bottom* row of this table.

	WHITE	BLACK
REMARRIED	_____%	_____%

b. What is the value of V?

V = _____

c. Is V statistically significant?

Yes   No

d. Blacks are less likely than whites to be remarried.

T   F

5. Could the relationship between race and remarriage actually be a by-product of differences in personal income? To find out, we will repeat the previous table using a control variable for the respondent's income.

> Data File: **GSS**
> Task: **Cross-tabulation**
> Row Variable: **8) DIV/REMAR.**
> Column Variable: **26) RACE**
> ➤ Control Variable: **29) R.INCOME**
> ➤ View: **Tables ($0K–17.4K)**
> ➤ Display: **Column %**

a. When only low-income respondents are considered, are blacks less likely to remarry?

Yes   No

b. Is V statistically significant?

Yes   No

Click on the arrow to advance to the table for the middle income category.

c. When only middle-income respondents are considered, are blacks less likely to remarry?

Yes   No

d. Is V statistically significant?

Yes   No

Click on the arrow to advance to the table for the high income category.

e. When only higher-income respondents are considered, are blacks less likely to remarry?

Yes   No

f. Is V statistically significant?

Yes   No

g. Which of the following statements is supported by these results? (circle one)

1. Blacks are less likely than whites to remarry regardless of social class.
2. Among those with low incomes, blacks are less likely than whites to remarry.
3. Among those with higher incomes, blacks are less likely than whites to remarry.

6. Does the size of the community in which people live affect their likelihood for remarriage? (Don't forget to clear the control variable from the previous question.)

> Data File: **GSS**
> Task: **Cross-tabulation**
> Row Variable: **8) DIV/REMAR.**
> ➤ Column Variable: **38) COMMUNITY**
> ➤ View: **Tables**
> ➤ Display: **Column %**

a. Fill in the percentaged results for the *bottom* row of this table.

	BIG CITY	SUBURB	SMALL TOWN	RURAL
REMARRIED	_____%	_____%	_____%	_____%

b. What is the value of V? V = _____

c. Is V statistically significant? Yes   No

d. Generally speaking, the larger the population of the community in which a person who is divorced lives: (circle one)

   1. the more likely the person is to be remarried.
   2. the less likely the person is to be remarried.
   3. does not influence his or her likelihood of being remarried.

e. How would you explain these results?

CHAPTER 15

FAMILIES AND EDUCATION: AN INTRODUCTION TO MULTIPLE REGRESSION

Tasks: Mapping, Correlation, Regression
Data Files: STATES, GLOBAL

Family and education have always been closely linked. Before 1642, when the Massachusetts colony passed the first mandatory school attendance law in the United States, children received most of their training in the home. The extent of the training children received, whether it was reading and writing or just the learning of a trade, varied by historical period, geography, and social class—but prior to industrialization, the idea of going to school was reserved for a privileged few.

In the United States, the formal shift of the responsibility for education occurred in 1954 in *Brown v. Board of Education*, when the Supreme Court ruled that education is a constitutional right that must be guaranteed by the state. So, what role does the family now play in the area of education? Who has more influence over a child's education—the family or the school? In 1966, sociologist James S. Coleman found that it was the quality and structure of the family, more than the quality and structure of the school, which determined the outcome of a child's education.[1] In the more than 30 years that have passed since Coleman's research, many changes have taken place both in schools and in the family. But as we shall see in this chapter, today's family is still very influential when it comes to education.

Let's start by looking at a map of average verbal SAT scores in the United States.

➤ *Data File:* **STATES**
➤ *Task:* **Mapping**
➤ *Variable 1:* **57) SAT.VERBAL**
➤ *View:* **Map**

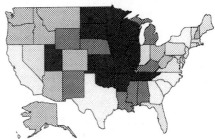

SAT.VERBAL -- 2000: VERBAL SCORE ON SAT (DES,2000)

Most of the states with the highest average scores on the verbal section of the SAT (Scholastic Aptitude Test) are located in the north central states. The lowest-scoring states are in the Southeast and along portions of each coast.

---

[1] Coleman, James S., et al. *Equality of Educational Opportunities* (Washington, D.C.: U.S. Government Printing Office, 1966).

Data File: **STATES**
Task: **Mapping**
Variable 1: **57) SAT.VERBAL**
➤ View: **List: Rank**

SAT.VERBAL: 2000: Verbal score on SAT

RANK	CASE NAME	VALUE
1	Iowa	589
2	North Dakota	588
3	South Dakota	587
4	Wisconsin	584
5	Minnesota	581
6	Kansas	574
7	Missouri	572
8	Utah	570
9	Illinois	568
10	Tennessee	563

As we look at the actual distribution we see that Iowa (589) has the highest average scores, followed by North Dakota (588) and South Dakota (587). The lowest scores are in South Carolina (484), Hawaii (488), and Georgia (488). Let's see if the distribution is similar for the scores on the math section of the SAT.

Data File: **STATES**
Task: **Mapping**
Variable 1: **57) SAT.VERBAL**
➤ Variable 2: **58) SAT.MATH**
➤ Views: **Map**

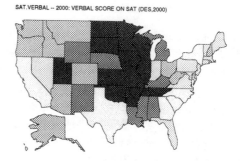

SAT.VERBAL -- 2000: VERBAL SCORE ON SAT (DES,2000)

r = 0.963**

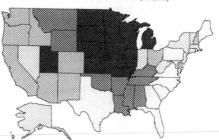

SAT.MATH -- 2000: MATHEMATICAL SCORE ON SAT (DES,2000)

The maps are strikingly similar. In fact, the correlation coefficient (r) is 0.963, which means the rankings are nearly identical.

Before explaining variations in SAT scores, it's important to examine the percentage of high school graduates in each state who take the SAT. Not all states encourage their schools to use the SAT. Some may recommend a different test; others may discourage standardized testing altogether.

Marriage and Family

Data File: **STATES**
Task: **Mapping**
Variable 1: **57) SAT.VERBAL**
➤ Variable 2: **59) %TAK.SAT**
➤ Views: **Map**

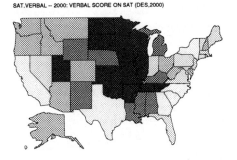

SAT.VERBAL -- 2000: VERBAL SCORE ON SAT (DES,2000)

r = −0.892**

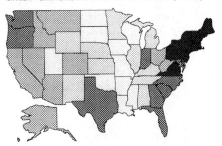

%TAK.SAT -- 2000: PERCENT OF GRADUATES TAKING THE SAT (DES,2000)

The states with the most students taking the exam are clearly not the states with the highest scores. The strong negative correlation (r = −0.892) tells us that the higher the percentage who take the test, the lower the average test score. It would appear that in those states that do not encourage its use, only the students who are likely to do well on the test are likely to take it. We will come back to the impact of this phenomenon in just a moment, but first look at some school-related variables that may affect SAT scores. Specifically, let's examine the pupil/teacher ratio, the amount of money spent per pupil, and teachers' salaries.

Data File: **STATES**
➤ Task: **Correlation**
➤ Variables: **57) SAT.VERBAL**
60) PUPIL/TCH
61) TEACHER$
62) $PER PUPIL

Correlation Coefficients
PAIRWISE deletion (1-tailed test)   Significance Levels: ** = .01, * = .05

	SAT.VERBAL	PUPIL/TCH	TEACHER$	$PER PUPIL
SAT.VERBAL	1.000 (50)	-0.038 (50)	-0.473 ** (50)	-0.393 ** (50)
PUPIL/TCH	-0.038 (50)	1.000 (50)	0.041 (50)	-0.348 ** (50)
TEACHER$	-0.473 ** (50)	0.041 (50)	1.000 (50)	0.820 ** (50)
$PER PUPIL	-0.393 ** (50)	-0.348 ** (50)	0.820 ** (50)	1.000 (50)

Two of the variables have a statistically significant relationship with verbal SAT scores—the amount of money spent per pupil (−0.393) and teachers' salaries (−0.473). But both correlations are negative! The more money the government spends on education, and the more pay teachers receive, the lower the SAT scores. What is going on here?

Think back to the map comparison of SAT scores with the percentage taking the SAT test. It was a negative correlation (r = −0.892); the more students who took the test, the lower the average score.

What if the states that spend the most on education are the same states where most of the students take the SAT? Then those states will have lower SAT scores, not because of spending, but because they have more students taking the exam.

Data File: **STATES**

Task: **Correlation**

➤ Variables: **59) %TAK.SAT**
61) TEACHER$
62) $PER PUPIL

Correlation Coefficients
PAIRWISE deletion (1-tailed test)   Significance Levels: ** = .01, * = .05

	%TAK.SAT	TEACHER$	$PER PUPIL
%TAK.SAT	1.000 (50)	0.610 ** (50)	0.607 ** (50)
TEACHER$	0.610 ** (50)	1.000 (50)	0.820 ** (50)
$PER PUPIL	0.607 ** (50)	0.820 ** (50)	1.000 (50)

Just as we might have suspected, there is a strong positive correlation between the percentage taking the SAT and spending per pupil (r = 0.607); the correlation between the percentage taking the SAT and teachers' salaries is also strong (r = 0.610). The states that spend the most money on education are the same ones that encourage taking the SAT. But how can we be sure which of these variables is actually having the negative impact on SAT scores? Maybe funding somehow does have a negative effect on school performance. Fortunately, social scientists have a technique for sorting out the impact of more than one variable at a time. It is called multiple regression.

Data File: **STATES**

➤ Task: **Regression**

➤ Dependent Variable: **57) SAT.VERBAL**

➤ Independent Variables: **62) $PER PUPIL**
61) TEACHER$
59) %TAKE.SAT

➤ View: **Graph (Summary)**

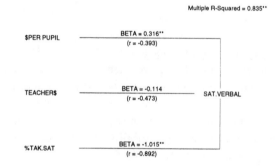

In the upper right corner, the screen reads: Multiple R–Squared = 0.835**. This stands for R^2 (pronounced "r squared"), which is a measure of the combined effects of the three independent variables on the dependent variable. Simply explained, this means that the variables reflecting spending per pupil, teachers' salaries, and the percentage of students taking the SAT together account for 83.5 percent of the variation in SAT scores (move the decimal point to the right two digits to convert it to a percentage). Put another way, if all states spent the same amount on education and had the same percentage of students taking the test, there would have been 83.5 percent less variation in the average SAT scores.

But that's not all we can see in this graphic. Beneath each of the horizontal lines is the value of r, which is Pearson's correlation coefficient. These are, of course, the same values as those shown in the set of correlations we have already examined. Above each line is the word "beta," followed by a numerical value. This stands for the standardized beta, which estimates **the independent effect of each independent variable on the dependent variable.**

The independent variables in this analysis are correlated with one another as well as with the dependent variable. What regression does is sort out the independent contributions of these three variables. Reading from the top of the graphic down, what we first discover is that spending per pupil actually has a positive effect on SAT scores. We know this because its beta is positive. A beta value that has one asterisk indicates significance at the .05 level; two asterisks indicates significance at the .01 level. Thus, the relationship between spending per pupil and SAT scores is statistically significant. Reading on down the graphic we find that teachers' salaries (beta = –0.114) does not have a statistically significant impact one way or the other. As we saw before, the percentage of students taking the SAT actually does have a negative relationship with average scores (beta = –1.015**). Unlike the case with the variables related to spending on education, the %TAK.SAT correlation by itself was fairly accurate.

The negative correlation between the salaries of teachers and SAT scores (r = –0.473) is an example of a spurious correlation. Spurious correlations occur between two variables because each is related to some unexamined additional variable or variables. When this other variable or set of variables is taken into account, spurious correlations disappear. In this case, the spurious correlation occurred because both TEACHER$ and SAT.VERBAL were related to the percentage of students taking the SAT. Once the variable %TAK.SAT was included in the analysis, the relationship between the salaries of teachers and SAT scores was no longer statistically significant–thus, it was a spurious, or false, correlation. The spurious correlation was caused by the fact that states which spend more on teacher salaries also have more students taking the SAT.

Let's use multiple regression to learn how some family-related variables affect SAT scores. According to Coleman, we would expect that states which have **more married couples should have higher SAT scores; in addition, states that have more families living below the poverty line should have lower SAT scores**. Since we have already learned that states with high percentages of students who take the SAT has a strong effect on SAT scores, we'll want to continue to control for this variable. So, select 59, or %TAK.SAT, as the third independent variable.

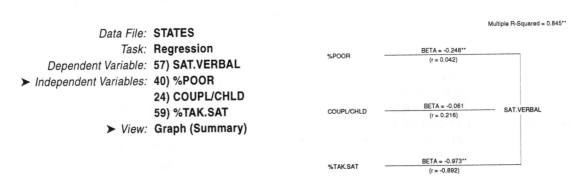

Data File: **STATES**
Task: **Regression**
Dependent Variable: **57) SAT.VERBAL**
➤ Independent Variables: **40) %POOR**
24) COUPL/CHLD
59) %TAK.SAT
➤ View: **Graph (Summary)**

Multiple R-Squared = 0.845**

%POOR — BETA = -0.248** (r = 0.042) — SAT.VERBAL

COUPL/CHLD — BETA = -0.061 (r = 0.216)

%TAK.SAT — BETA = -0.973** (r = -0.892)

For this example, the easiest way to select a new set of variables is to first click the [Clear All] button.

Our hypothesis is partially supported. The percentage of families living below the poverty line is related to SAT scores (beta = –0.248**); the more poor families there are, the lower the average SAT score. However, the percentage of married couples does not appear to affect SAT scores. Although there is a weak correlation between the two variables (r = 0.216), the relationship virtually disappears when the impact of the other variables is included (beta = –0.061). This is another example of a spurious correlation. At least at the state level, the percentage of poor families in a state is a better predictor of that state's SAT scores than is the number of two-parent households. The fact that a state has more two-parent families does not guarantee that it will have higher average scores on the SAT.

Let's explore this point a little further by focusing on single-parent families. Does either the number of divorces or the teenage birth rate affect SAT scores?

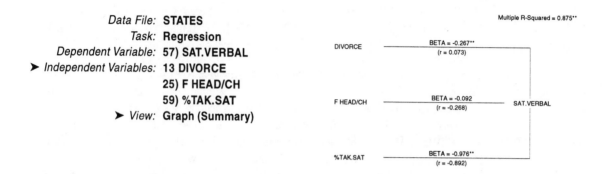

> *Data File:* **STATES**
> *Task:* **Regression**
> *Dependent Variable:* **57) SAT.VERBAL**
> ➤ *Independent Variables:* **13 DIVORCE**
> **25) F HEAD/CH**
> **59) %TAK.SAT**
> ➤ *View:* **Graph (Summary)**

The results show that the percentage of female-headed households with children is not related to SAT scores in a statistically significant manner (beta = –0.092); however, the divorce rate has a statistically significant effect (beta = –0.267**). The higher the divorce rate, the lower the SAT scores. Using state-level data, we cannot tell whether those who took the SAT lived in female-headed families or in families going through a divorce. But we can see that a state's divorce rate in a given year has more of an effect on SAT scores than does the percentage of female-headed families.

So far, the dependent variable has been scores on the SAT, which is a test most students take as they are completing high school. Would these findings be different if we looked at younger children—for example, eighth graders? Let's see how some of these same variables affect the scores on a math proficiency test given to students in the eighth grade.

> *Data File:* **STATES**
> *Task:* **Regression**
> ➤ *Dependent Variable:* **64) MATH SCORE**
> ➤ *Independent Variables:* **13) DIVORCE**
> **25) F HEAD/CH**
> **40) %POOR**
> ➤ *View:* **Graph (Summary)**

The results here are quite different from those we saw when we used SAT scores as the dependent variable. Here, both the divorce rate (beta = –0.487**) and the percentage of female-headed homes (beta = –0.607**) are statistically significant. However, the percentage of families below the poverty line (beta = –0.019) does not have a statistically significant impact on math proficiency scores. Together, these variables account for about two-thirds of the variation in math proficiency scores (Multiple R-Squared = 0.648**).

Marriage and Family

Again, we cannot assume that the students with the lowest math scores are necessarily living in single-parent families, but eighth-grade math scores in states that have high rates of female-headed homes do tend to be lower overall.

One of the most difficult aspects of being a single parent is task overload—there is simply too much to do and not enough time. Work schedules and maintaining a household single-handedly can cut into time that single parents may want to devote to their children. Of course, children in many two-parent households are unsupervised to the same extent. But it is easier for parents in two-parent households to find time to devote to parenting. One indirect measure of parental interaction in a community is the amount of time the kids who live there spend watching television. Communities in which the children are supervised less will generally watch more television. By including this variable in a regression, along with measures of family structure, we can separate the effects of supervision from other aspects of family structure.

<table>
<tr><td align="right">Data File:</td><td>STATES</td></tr>
<tr><td align="right">Task:</td><td>Regression</td></tr>
<tr><td align="right">Dependent Variable:</td><td>64) MATH SCORE</td></tr>
<tr><td align="right">➤ Independent Variables:</td><td>25) F HEAD/CH</td></tr>
<tr><td></td><td>49) TV>6HRS</td></tr>
<tr><td align="right">➤ View:</td><td>Graph (Summary)</td></tr>
</table>

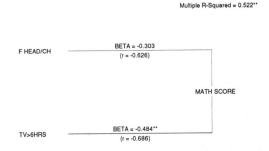

Multiple R-Squared = 0.522**

The percentage of children who watch more than six hours of television per day does indeed have a statistically significant negative effect on math scores (beta = –0.484**). In fact, this variable explains most of the effect of single-parent households on math scores (beta = –0.303). Communities in which children watch more television tend to have lower math scores overall.

What we have seen is that the relationship among schools, family, and academic achievement is very complex. There are no simple answers as to why some children do better in school than others, and there are no quick avenues to ensuring academic success. In the worksheet section that follows, you will have the opportunity to use multiple regression to examine some other education- and family-related issues that can be equally challenging to families and to society.

WORKSHEET

15

NAME:

COURSE:

DATE:

Workbook exercises and software are copyrighted. Copying is prohibited by law.

REVIEW QUESTIONS

Based on the first part of this chapter, answer True or False to the following items:

Most of the states with the highest SAT scores are located in New England and along the west coast.	T	F
States with more students taking the SAT tend to have higher average scores.	T	F
According to the regression analysis, states that spend more per pupil on education tend to have higher SAT scores.	T	F
States with a greater proportion of poor families tend to have lower SAT scores.	T	F
States with higher divorce rates tend to have lower math scores.	T	F

EXPLORIT QUESTIONS

1. Let's look at the relationship among four school-related variables: high school graduation rates, pupil/teacher ratios, spending per pupil, and teachers' salaries.

> ➤ *Data File:* **STATES**
> ➤ *Task:* **Correlation**
> ➤ *Variables:* **55) HS GRAD**
> **60) PUPIL/TCH**
> **61) TEACHER$**
> **62) $PER PUPIL**

a. What is the correlation between pupil/teacher ratios and graduation rates? _____

Is the correlation statistically significant? Yes No

b. What is the correlation between teachers' salaries and graduation rates? _____

Is the correlation statistically significant? Yes No

c. What is the correlation between spending per pupil and graduation rates? _____

Is the correlation statistically significant? Yes No

Chapter 15: Families and Education: An Introduction to Multiple Regression　　　　　245

d. States with high pupil/teacher ratios in their schools tend to have high dropout rates.   T F

e. States that spend more per pupil tend to have high dropout rates.   T F

f. States with the highest salaries for teachers tend to have high graduation rates.   T F

2. What are the factors that influence the pupil/teacher ratio? Does the urbanity of a state or median family income have an effect? To answer this question, let's look at the following regression.

> Data File: **STATES**
> ➤ Task: **Regression**
> ➤ Dependent Variable: **60) PUPIL/TCH**
> ➤ Independent Variables: **3) %URBAN**
>   **62) $PER PUPIL**
>   **38) MED.FAM$**
> ➤ View: **Graph (Summary)**

a. What is the explained variance (the combined effect) of these three independent variables (R-Squared)? (Remember, convert your decimal value to a percentage.) _____%

b. Is the beta for the relationship between PUPIL/TCH and %URBAN statistically significant?   Yes   No

c. Is the beta for the relationship between PUPIL/TCH and $PER PUPIL statistically significant?   Yes   No

d. Is the beta for the relationship between PUPIL/TCH and MED.FAM$ statistically significant?   Yes   No

e. Based on the results, we would expect to find more pupils per teacher in states with (circle one)
   1. a large rural population.
   2. a large urban population.

f. We would expect to find fewer pupils per teacher in states (circle one)
   1. that spend more per pupil.
   2. that spend less per pupil.
   3. with high median family incomes.

3. Let's look at two other factors—two-parent homes and the availability of libraries—that might influence the high school graduation rate.

> Data File: **STATES**
> Task: **Regression**
> ➤ Dependent Variable: **55) HS GRAD**
> ➤ Independent Variables: **24) COUPL/CHLD**
> **63) LIBRARIES**
> ➤ View: **Graph (Summary)**

a. What is the explained variance (the combined effect) of these two independent variables (R-Squared)? (Remember, convert your decimal value to a percentage.) _____%

b. Is the beta for the relationship between HS GRAD and COUPL/CHLD statistically significant?   Yes   No

c. Is the beta for the relationship between HS GRAD and LIBRARIES statistically significant?   Yes   No

d. Based on the results, high school graduation rates are influenced by (circle the best answer)
   1. the proportion of two-parent households.
   2. the availability of libraries.
   3. both the proportion of two-parent households and the availability of libraries.
   4. neither the proportion of two-parent households nor the availability of libraries.

4. States with high percentages of African Americans also have lower high school graduation rates. In previous chapters we have found that there is a higher proportion of female-headed families among African Americans. So, let's use both %BLACK and F HEAD/CH as independent variables to see how each of these factors influences the high school graduation rate.

> Data File: **STATES**
> Task: **Regression**
> Dependent Variable: **55) HS GRAD**
> ➤ Independent Variables: **35) %BLACK**
> **25) F HEAD/CH**
> ➤ View: **Graph (Summary)**

a. What is Pearson's correlation (r) between %BLACK and HS GRAD?   r = _____

b. What is the independent effect (beta) of %BLACK?   beta = _____

c. Is this beta statistically significant?   Yes   No

d. There is a strong positive correlation between %BLACK and HS GRAD. However, when you control for female-headed families the beta is not statistically significant.   T   F

e. These results suggest that the relationship between %BLACK and HS GRAD
is spurious.   T   F

5. Let's look at the dropout rates of young children using the GLOBAL data file. The variable 29) % GO 5TH
is the percentage of children who reach grade 5 before quitting.

> *Data File:* **GLOBAL**
> *Task:* **Correlation**
> *Variables:* **29) % GO 5TH**
>   **4) BIRTH RATE**
>   **22) $ PER CAP**

a. Fill in the correlation results from the top line of the table.

	BIRTH RATE	$ PER CAP
% GO 5TH	_____	_____

Now go to the REGRESSION task and make 29) % GO 5TH the dependent variable.

> *Data File:* **GLOBAL**
> *Task:* **Regression**
> *Dependent Variable:* **29) % GO 5TH**
> *Independent Variables:* **4) BIRTH RATE**
>   **22) $ PER CAP**
> *View:* **Graph (Summary)**

b. What is the explained variance (the combined effect) of these two independent
variables?   _____%

c. What is the independent effect (beta) of BIRTH RATE?   beta = _____

d. What is the independent effect (beta) of $ PER CAP?   beta = _____

e. Are both betas and the R-SQ values statistically significant?   Yes   No

f. Compare the correlation and the beta for $ PER CAP and % GO 5TH. Would
it be accurate to say that when you control for the birth rate, the relationship
between gross national product per capita and the dropout rates of children
is greatly weakened?   Yes   No

g. Compare the correlation and the beta for BIRTH RATE and % GO 5TH. Would
it be accurate to say that when you control for the gross national product per
capita, the relationship between birth rates and the dropout rates of children
is greatly weakened?   Yes   No

APPENDIX: VARIABLE NAMES AND SOURCES

Note for MicroCase Users: These data files may be used with the MicroCase Analysis System. If you are moving variables from these files into other MicroCase files, or vice versa, you may need to reorder the cases. Also note that files that have been modified in MicroCase will not function in Student ExplorIt.

◆ DATA FILE: CULTURES ◆

1) CASE NAME
2) REGION
3) AG.DEVELOP
4) %FEM.SUBST
5) %FEM.SUBS*
6) %FEM.GATHR
7) %FEM.AGRIC
8) EXT.FAMILY
9) MONOGAMY?
10) MAR.FORM
11) MARRIAGE $
12) COUPLE PIC
13) GROOM PICK
14) MEN PICK
15) MEN PICK2
16) FEM.PICK
17) LEVIRATE
18) PATRILOCAL
19) COUPLE BED
20) PRIVATE
21) SPOUSE EAT
22) PLAY MATES
23) REMARRIAGE
24) M-DIVORCE?
25) MALE DIV.

26) DIVORCE RT
27) DIV COMMON
28) GROUNDS? M
29) GROUNDS? F
30) INFANTCARE
31) CHILD CARE
32) WARM:KIDS
33) HIT KIDS
34) HIT KIDS*
35) C.NEGLECT
36) SUPERVISE?
37) WANT BOYS?
38) COMPETE B.
39) COMPETE G.
40) SELF-REL B
41) SELF-REL.G
42) TOUGH BOYS
43) TUFF BOYS*
44) TOUGH GIRL
45) TUFF GIRL*
46) OLD GROOMS
47) HOUSEWORK?
48) WIFE DEFER
49) BEAT WIVES
50) OVER WIVES

51) F.INFERIOR
52) FEM.PROPTY
53) F.KIN POWR
54) FEM.LIVES?
55) FEM.SOLID.
56) SEX SEGR.
57) HEIR'S SEX
58) SEX (BOYS)
59) SEX(BOYS)*
60) SEX(GIRLS)
61) SEX(GIRL)*
62) FEM.SEXUAL
63) PREMARITAL
64) EXTRAMARTL
65) FEM EXTRA
66) CHAPERONES
67) TEEN SEX?
68) ANTI GAY
69) HOMOSEXUAL
70) RAPE
71) MALE LEADR
72) SEX/POLITY
73) SUBS/MODE
74) FIXITY
75) MOBILITY

✦ DATA FILE: GLOBAL ✦

1) COUNTRY	26) LITERACY	51) POL RIGHTS
2) URBAN %	27) PRMSC/CP	52) CIVIL LIBL
3) POP GROWTH	28) UNIV/CP	53) %MUSLIM
4) BIRTH RATE	29) % GO 5TH	54) %CHRISTIAN
5) FERTILITY	30) FEM.PROF.	55) %CATHOLIC
6) LARGE FAML	31) %FEM.LEGIS	56) %HINDU
7) INF. MORTL	32) %FEM.WRK	57) %BUDDHIST
8) MOM MORTAL	33) M/F EDUC.	58) REL.PERSON
9) CONTRACEPT	34) GENDER EQ	59) CH.ATTEND
10) ABORTION	35) FEM POWER	60) GOD IMPORT
11) ABORT LEGL	36) SEX MUTIL	61) PRAY?
12) MOM HEALTH	37) SINGLE MOM	62) EX-MARITAL
13) AB. UNWANT	38) WORK MOM?	63) MINOR SEX
14) DEATH RATE	39) HOME&KIDS	64) GAY SEX
15) LIFEX MALE	40) CHORES?	65) ANTI-GAY
16) LIFEX FEM	41) HOME LIFE?	66) PROSTITUTE
17) SEX RATIO	42) SPOUSE SEX	67) AIDS
18) ECON DEVEL	43) HAPPY SEX?	68) SUICIDE
19) THREEWORLD	44) FAMILY IMP	69) ALCOHOL
20) QUAL. LIFE	45) KID MANNER	70) REGION
21) INEQUALITY	46) KID INDEPN	71) AS/AFR/LAT
22) $ PER CAP	47) KID OBEY	72) EUROPE
23) % AGRIC $	48) WED PASSE'	73) F/M EMPLOY
24) % INDUS $	49) VERY HAPPY	74) %FEM.HEADS
25) PUB EDUCAT	50) GOVERNMENT	

✦ DATA FILE: GSS ✦

1) MARITAL	19) ROMANCE	37) REGION
2) MARRIED?	20) WILL WED1	38) COMMUNITY
3) EVER MAR?	21) WILL WED2	39) PARTY
4) DIVORCED?	22) LIVE WITH	40) POL. VIEW
5) WIDOWED?	23) PARTNER	41) # SIBS
6) DIV/MAR	24) AGE	42) DAD EDUC.
7) SINGLE/MAR	25) SEX	43) MOM EDUC.
8) DIV/REMAR.	26) RACE	44) MATE EDUC.
9) REMARRIAGE	27) EDUCATION	45) FAMILY @16
10) SNG.PARENT	28) INCOME	46) MA WRK GRW
11) FAM.STAGE	29) R.INCOME	47) MOVERS
12) SING.W/KID	30) WORK STAT	48) RELIGION
13) # CHILDREN	31) FAM.WORK	49) R.FUND/LIB
14) IDEAL#KIDS	32) MAR.F.WRK	50) ATTEND
15) HH SIZE	33) WORK HOURS	51) R.FUND@16
16) KIDS<18	34) OCCUPATION	52) MOMS RELIG
17) KIDS<6	35) CHANGE $?	53) POPS RELIG
18) AGE KD BRN	36) JOB SATIS.	54) SPOUS.RELG

◆ DATA FILE: GSS (CONT.) ◆

55) MA ATTEND	75) DIV.EASY?	95) ACQNT SEX?
56) PA ATTEND	76) HEALTH	96) PIKUP SEX?
57) INTERMAR.?	77) LIFE	97) PAID SEX?
58) MARRY BLK	78) SOC.KIN	98) SEX OF SEX
59) FAMWHTS	79) SOC.NEIGH.	99) SEX FREQ.
60) FAMBLKS	80) SOC.FRIEND	100) EV.PAIDSEX
61) FAMJEWS	81) SOC. BAR	101) EVER STRAY
62) FAMHSPS	82) WATCH TV	102) CONDOM
63) FAMASNS	83) WEB HOURS	103) RELATE SEX
64) MOTH.WORK?	84) POP MUSIC	104) COHABITATE
65) PRESCH.WRK	85) EAT OUT	105) COHAB.FRST
66) WIFE@HOME	86) SEE FILM	106) ABORT.DEF.
67) WOMEN WORK	87) SEX ED?	107) ABORT.HLTH
68) MARR.ROLES	88) PREM.SEX	108) ABORT RAPE
69) DOMES.DUTY	89) TEEN SEX	109) ABORT NO$
70) DISC MAN	90) XMAR.SEX	110) ABORT SING
71) DISC WOMAN	91) HOMO.SEX	111) ABORT ANY
72) HIRE WOMEN	92) X-MOVIE?	112) SPANKING
73) HAPPY?	93) SX.PRTNRS?	113) KID OBEY
74) HAP.MARR.?	94) FRIEND SEX	114) THINK SELF

◆ DATA FILE: STATES ◆

1) STATE NAME	25) F HEAD/CH	49) TV>6HRS
2) WARM WINTR	26) M HEAD/CH	50) PLAYBOY
3) %URBAN	27) FERTILITY	51) V.CRIME
4) $RURAL	28) SMALLB/W	52) P.CRIME
5) NO MOVE	29) UNWED	53) MURDER
6) POP.GROW	30) TEENMOM	54) RAPE
7) %MALE	31) ABORTION	55) HS GRAD
8) %FEMALE	32) INF.MORTAL	56) COLLEGE
9) SEX RATIO	33) AIDS	57) SAT.VERBAL
10) %UNDER 18	34) %WHITE	58) SAT.MATH
11) %OVER 65	35) %BLACK	59) %TAK.SAT
12) MARRIAGE	36) %ASIAN	60) PUPIL/TCH
13) DIVORCE	37) %HISPANIC	61) TEACHER$
14) %DIVORCED	38) MED.FAM$	62) $PER PUPIL
15) %SINGLE M	39) PER CAP$	63) LIBRARIES
16) %MAR.MEN	40) %POOR	64) MATH SCORE
17) %DIV.MEN	41) %FEM.WRK	65) %GWBUSH
18) %SINGLE F	42) %NO.RELIG	66) %GORE
19) %MAR.FEM	43) % JEWISH	67) %NADER
20) %DIV.FEM	44) % CATHOLIC	68) %FEM.WRK40
21) %WIDOWS	45) % BAPTIST	69) MEDIAN$ 40
22) %MARRIED	46) CH.MEMBER	70) F WAGES 20
23) %SINGLE	47) SUICIDE	71) $FEM.WRK20
24) COUPL/CHLD	48) RITALIN	72) DIVORCE 22

✦ DATA FILE: TRENDS ✦

1) Date
2) YEAR
3) POPULATION
4) MARR. RATE
5) DIV. RATE
6) SINGLE%
7) MARRIED%
8) DIVORCED%
9) EVER DIV%
10) REMARRIED%
11) DURATION

12) M AGE@MAR
13) F AGE@MAR
14) BIRTH RATE
15) INF. DEATH
16) NO KIDS%
17) KIDS 3+%
18) HAP.MAR.?%
19) SOC.KIN%
20) NO MOVE%
21) XMAR.SEX%
22) PREM.SEX%

23) %NO HS
24) %HS
25) %COLLEGE
26) %FEM EMPL
27) M FEM EMP%
28) HOMEMAKER%
29) WOMEN WORK
30) HOMO.SEX
31) INTERMAR.?
32) TEEN BIRTH

SOURCES

CULTURES

The CULTURES data file is based on the Standard Cross-Cultural Sample by George Peter Murdock and Douglas R. White (1969). This sample utilizes a subset of 186 preindustrial societies from the Ethnographic Atlas (1971, Ethnology).

GLOBAL

The data in the GLOBAL file are from a variety of sources. The variable description for each variable uses the following abbreviations to indicate the source.

FITW: Freedom in the World, 1997, Freedom House

HDR: Human Development Report, 1994 and 1998, United Nations Development Program

IP: International Profile: Alcohol and Other Drugs, 1994, Alcoholism and Drug Addiction Research Foundation

NBWR: The New Book of World Rankings, 3d Edition, 1991, Facts on File

PON: The Progress of Nations, 1996, UNICEF

SAUS: Statistical Abstract of the United States, 1998, U.S. Department of Commerce

TWF: The World Factbook, 1997, Central Intelligence Agency

TWW: The World's Women, 1995, United Nations

UNSY: United Nations Statistical Yearbook, 1997, United Nations

WCE: World Christian Encyclopedia, David R. Barrett, editor, Oxford University Press, 1982

WDI: World Development Indicators, 1998, published annually by the World Bank

WVS: World Values Survey, 1981–1984, 1990–1993, Institute for Social Research, Inter-University Consortium for Political and Social Research

GSS

The GSS data file is based on selected variables from the National Opinion Research Center (University of Chicago) General Social Survey for 2000, distributed by the Roper Center and the Inter-University Consortium for Political and Social Research. The principal investigators are James A. Davis and Tom W Smith.

STATES

The data in the STATES file are from a variety of sources. The variable description for each variable uses the following abbreviations to indicate the source.

ABC: Blue Book, Audit Bureau of Circulation

CDC: Center for Disease Control

CENSUS: The summary volumes of the U.S. Census

DEA: Drug Enforcement Administration

DES: Digest of Education Statistics, U.S. Dept. of Education

FEC: Federal Election Commission

HCSR: Health Care State Rankings, Morgan Quitno

KOSMIN: Kosmin, Barry A. 1991 Research Report: The National Survey of Religious Identification, New York: CUNY Graduate Center

NAEP: National Assessment of Educational Progress, U.S. Dept. of Education

SA: Statistical Abstract of the United States

UCR: The Uniform Crime Reports, U.S. Dept. of Justice

US Bureau of the Census, Report ST-99-09

US Bureau of the Census, Report ST-96-2

US Bureau of the Census, Report ST-96-11

U.S. FISH & WILDLIFE: Data provided by U.S. Fish and Wildlife Service

USA Today, 8/6/90 (cites the Senate Judiciary Committee)

Variables with no source shown are from U.S. Census publications.

LICENSE AGREEMENT FOR WADSWORTH GROUP, A DIVISION OF THOMSON LEARNING, INC.

You the customer, and Wadsworth Group incur certain benefits, rights, and obligations to each other when you open this package and use the materials it contains. BE SURE TO READ THE LICENSE AGREEMENT CAREFULLY, SINCE BY USING THE SOFTWARE YOU INDICATE YOU HAVE READ, UNDERSTOOD, AND ACCEPTED THE TERMS OF THIS AGREEMENT.

Your rights

1. You enjoy a non-exclusive license to use the enclosed materials on a single computer that is not part of a network or multi-machine system in consideration of the payment of the required license fee, (which may be included in the purchase price of an accompanying print component), and your acceptance of the terms and conditions of this agreement.

2. You own the disk on which the program/data is recorded, but you acknowledge that you do not own the program/data recorded on the disk. You also acknowledge that the program/data is furnished "AS IS," and contains copyrighted and/or proprietary and confidential information of Wadsworth Group, a division of Thomson Learning, Inc.

3. If you do not accept the terms of this license agreement you must not install the disk and you must return the disk within 30 days of receipt with proof of payment to Wadsworth Group for full credit or refund.

These are limitations on your rights

1. You may not copy or print the program/data for any reason whatsoever, except to install it on a hard drive on a single computer, unless copying or printing is expressly permitted in writing or statements recorded on the disk.

2. You may not revise, translate, convert, disassemble, or otherwise reverse engineer the program/data.

3. You may not sell, license, rent, loan, or otherwise distribute or network the program/data.

4. You may not export or re-export the disk, or any component thereof, without the appropriate U.S. or foreign government licenses. Should you fail to abide by the terms of this license or otherwise violate Wadsworth Group's rights, your license to use it will become invalid. You agree to destroy the disk immediately after receiving notice of Wadsworth Group's termination of this agreement for violation of its provisions.

U.S. Government Restricted Rights

The enclosed multimedia, software, and associated documentation are provided with RESTRICTED RIGHTS. Use, duplication, or disclosure by the Government is subject to restrictions as set forth in subdivision (c)(1)(ii) of the Rights in Technical Data and Computer Software clause at DFARS 252.277.7013 for DoD contracts, paragraphs (c) (1) and (2) of the Commercial Computer Software-Restricted Rights clause in the FAR (48 CFR 52.227-19) for civilian agencies, or in other comparable agency clauses. The proprietor of the enclosed multimedia, software, and associated documentation is Wadsworth Group, 10 Davis Drive, Belmont, California 94002.

Limited Warranty

Wadsworth Group also warrants that the optical media on which the Product is distributed is free from defects in materials and workmanship under normal use. Wadsworth Group will replace defective media at no charge, provided you return the Product to Wadsworth Group within 90 days of delivery to you as evidenced by a copy of your invoice. If failure of disc(s) has resulted from accident, abuse, or misapplication, Wadsworth Group shall have no responsibility to replace the disc(s). THESE ARE YOUR SOLE REMEDIES FOR ANY BREACH OF WARRANTY.

EXCEPT AS SPECIFICALLY PROVIDED ABOVE, WADSWORTH GROUP, A DIVISION OF THOMSON LEARNING, INC. AND THE THIRD PARTY SUPPLIERS MAKE NO WARRANTY OR REPRESENTATION, EITHER EXPRESSED OR IMPLIED, WITH RESPECT TO THE PRODUCT, INCLUDING ITS QUALITY, PERFORMANCE, MERCHANTABILITY, OR FITNESS FOR A PARTICULAR PURPOSE. The product is not a substitute for human judgment. Because the software is inherently complex and may not be completely free of errors, you are advised to validate your work. IN NO EVENT WILL WADSWORTH GROUP OR ANY THIRD PARTY SUPPLIERS BE LIABLE FOR DIRECT, INDIRECT, SPECIAL, INCIDENTAL, OR CONSEQUENTIAL DAMAGES ARISING OUT OF THE USE OR INABILITY TO USE THE PRODUCT OR DOCUMENTATION, even if advised of the possibility of such damages. Specifically, Wadsworth Group is not responsible for any costs including, but not limited to, those incurred as a result of lost profits or revenue, loss of use of the computer program, loss of data, the costs of recovering such programs or data, the cost of any substitute program, claims by third parties, or for other similar costs. In no case shall Wadsworth Group's liability exceed the amount of the license fee paid. THE WARRANTY AND REMEDIES SET FORTH ABOVE ARE EXCLUSIVE AND IN LIEU OF ALL OTHERS, ORAL OR WRITTEN, EXPRESS OR IMPLIED. Some states do not allow the exclusion or limitation of implied warranties or limitation of liability for incidental or consequential damage, so that the above limitations or exclusion may not apply to you.

This license is the entire agreement between you and Wadsworth Group and it shall be interpreted and enforced under California law. Should you have any questions concerning this License Agreement, write to Technology Department, Wadsworth Group, 10 Davis Drive, Belmont, California 94002.